SpringerBriefs in Molecular Science

Chemistry of Foods

Series editor

Salvatore Parisi, Industrial Consultant, Palermo, Italy

The series Springer Briefs in Molecular Science: Chemistry of Foods presents compact topical volumes in the area of food chemistry. The series has a clear focus on the chemistry and chemical aspects of foods, topics such as the physics or biology of foods are not part of its scope. The Briefs volumes in the series aim at presenting chemical background information or an introduction and clear-cut overview on the chemistry related to specific topics in this area. Typical topics thus include: - Compound classes in foods – their chemistry and properties with respect to the foods (e.g. sugars, proteins, fats, minerals, …) - Contaminants and additives in foods – their chemistry and chemical transformations - Chemical analysis and monitoring of foods - Chemical transformations in foods, evolution and alterations of chemicals in foods, interactions between food and its packaging materials, chemical aspects of the food production processes - Chemistry and the food industry – from safety protocols to modern food production The treated subjects will particularly appeal to professionals and researchers concerned with food chemistry. Many volume topics address professionals and current problems in the food industry, but will also be interesting for readers generally concerned with the chemistry of foods. With the unique format and character of Springer Briefs (50 to 125 pages), the volumes are compact and easily digestible. Briefs allow authors to present their ideas and readers to absorb them with minimal time investment. Briefs will be published as part of Springer's eBook collection, with millions of users worldwide. In addition, Briefs will be available for individual print and electronic purchase. Briefs are characterized by fast, global electronic dissemination, standard publishing contracts, easy-to-use manuscript preparation and formatting guidelines, and expedited production schedules. Both solicited and unsolicited manuscripts focusing on food chemistry are considered for publication in this series.

More information about this series at http://www.springer.com/series/11853

Marcella Barbera · Giovanni Gurnari

Wastewater Treatment and Reuse in the Food Industry

 Springer

Marcella Barbera
ARPA
Regional Environmental Protection Agency
Ragusa
Italy

Giovanni Gurnari
Benaquam S.R.L.
Dogana
Republic of San Marino

ISSN 2191-5407   ISSN 2191-5415 (electronic)
SpringerBriefs in Molecular Science
ISSN 2199-689X   ISSN 2199-7209 (electronic)
Chemistry of Foods
ISBN 978-3-319-68441-3   ISBN 978-3-319-68442-0 (eBook)
https://doi.org/10.1007/978-3-319-68442-0

Library of Congress Control Number: 2017955234

© The Author(s) 2018
This work is subject to copyright. All rights are reserved by the Publisher, whether the whole or part of the material is concerned, specifically the rights of translation, reprinting, reuse of illustrations, recitation, broadcasting, reproduction on microfilms or in any other physical way, and transmission or information storage and retrieval, electronic adaptation, computer software, or by similar or dissimilar methodology now known or hereafter developed.
The use of general descriptive names, registered names, trademarks, service marks, etc. in this publication does not imply, even in the absence of a specific statement, that such names are exempt from the relevant protective laws and regulations and therefore free for general use.
The publisher, the authors and the editors are safe to assume that the advice and information in this book are believed to be true and accurate at the date of publication. Neither the publisher nor the authors or the editors give a warranty, express or implied, with respect to the material contained herein or for any errors or omissions that may have been made. The publisher remains neutral with regard to jurisdictional claims in published maps and institutional affiliations.

Printed on acid-free paper

This Springer imprint is published by Springer Nature
The registered company is Springer International Publishing AG
The registered company address is: Gewerbestrasse 11, 6330 Cham, Switzerland

Contents

Chapter 1
Water Reuse in the Food Industry: Quality of Original Wastewater Before Treatments

Abstract This chapter introduces one of the most important emergencies in the world of food and non-food industries: the availability of clean and drinking water. Water use has more than tripled globally since 1950: water quality and its scarcity are increasingly recognised as one of the most important environmental threats to humankind. In addition, the food and beverage processing industry requires copious amounts of water. For these reasons, direct and indirect water reuse systems are becoming more and more interesting and promising technologies. Different reuse guidelines have been recently issued as the result of risk assessment and management approaches linked to health-based targets. Chemical and biological features of wastewaters originated from different food processing environments have to be carefully analysed and adequate countermeasures have to be taken on these bases in relation to the specific food processing activity.

Keywords BOD/COD ratio · Fertiliser · Pesticide · Risk assessment · Suspended solids · Wastewater · Water reuse

Abbreviations

BOD	Biochemical oxygen demand
COD	Chemical oxygen demand
FAO	Food and Agriculture Organization
HACCP	Hazard Analysis and Critical Control Point
SWW	Slaughterhouses Wastewater
TSS	Total suspended solids
UNICEF	United Nations International Children's Emergency Fund
US EPA	United States Environmental Protection Agency
USA	United States of America
WHO	World Health Organization

© The Author(s) 2018

M. Barbera and G. Gurnari, *Wastewater Treatment and Reuse in the Food Industry*, Chemistry of Foods, https://doi.org/10.1007/978-3-319-68442-0_1

1.1   Food Industry and Generated Industrial Effluents: An Overview

In most industrial processes, water is the most extensively used raw material for the production of high-value products. Water use has more than tripled globally since 1950, and one out of every six persons does not have regular access to safe drinking water. At present, more than 700 million people worldwide lack access to safe water and sanitation affects the health of 1.2 billion people annually [1]. Water quality and its scarcity are increasingly recognised as one of the most important environmental threats to humankind [2]. In addition, steady economic development, particularly in emerging market economies, has translated into demand for a more varied diet, including meat and dairy products, putting additional pressure on water resources [3]. The food and beverage processing industry requires copious amounts of water; actually, this sector is the third largest industrial user of water [4]. In general, 75% of water used is considered useful because of its drinking quality in the food and beverage industrial sector as a whole [5]. More than two-thirds of all freshwater abstraction worldwide (and up to 90% in some countries) go towards food production: freshwater resources are depleted in many areas of the world. Some estimation reports that 35% of the world's population will live in countries affected by water stress or scarcity by 2025 [6]. Therefore, the food industry must address the future trends relating to this resource, in common with other industries, and move towards increasing efficiency in water use. Water consumption in the production and treatment of food and drink industry varies depending on different factors, such as the diversity of each manufacturing subsector, the number of end products, the capacity of the plant, the type of applied processes, employed equipment, automation levels, systems used for cleaning operations. [7, 8]. Wastewater resulting from food industries, including the agro-industrial sector, is obtained as one of the final products of human activities, which are associated with processing, manufacturing and raw material handlings, generated from medium- to large-scale industries. This wastewater arises from cooling, heating, extraction, reaction of by-products, washing and quality control as a result of specification products being rejected. The characteristic of these effluents depends on the quality of water used by different types of industries, as well as the community and treatment of such wastewater [9]. Industrial wastewater is difficult to characterise as it varies according to processes, season and related products [10]. Generally, the main contaminants are microorganisms, biodegradable organic material, sanitising products, fertilisers, pesticides, metals, nutrients, organic and inorganic materials.

1.2   Water and Wastewater Reutilisation

In urban areas, demand for water has been increasing steadily, owing to population growth, industrial development and expansion of irrigated peri-urban agriculture. As a consequence, an increment of the pollution of freshwater can be observed due to the inadequate discharge of wastewater, especially in developing countries [11]. An increase in industrial activities, along with the discharge of high-strength wastewater from various industries, results in challenges with regard to methods that are used to remediate contaminants in the water in order to limit its environmental impact.

At present, water management is conducting improper depletion of water resources of surface and groundwater. For these reasons, reduced water availability is already leading to attempts by the food industry to optimise its use. Reuse of water in the food industry is extremely interesting due to the increasing cost of water and water discharge and its treatment. Wastewater reuse potential in different industries depends on waste volume, concentration and characteristics, best available treatment technologies, operation and maintenance costs, availability of raw water and effluent standards. Radical changes in industrial wastewater reuse have to take into consideration rapidly depleting resources, environmental degradation, public attitude and health risks to workers and consumers.

Water quality requirements are a function of the type of food, processing conditions and methods of final preparation in the home (cooked/uncooked products) [12]. Water and wastewater reutilisation, costs of treatment and disposal guidelines remain the most critical factors for the development of sustainable water use for food and beverage industries, especially if access to water resources is required continually and with notable amounts. Consequently, there is an urgent need to improve the efficiency of water consumption and to augment the existing sources of water with more sustainable alternatives.

There are modern and traditional approaches for efficiency improvements and augmentation [7]. The move towards wastewater reuse is reflected in different cleaner production approaches such as internal recycling, reuse of treated industrial or municipal wastewater and reuse of treated effluents for other activities. Reusing wastewater is an attractive economic alternative, and it can be a useful strategy when speaking of essential preservation for future generations. A cautious use also reduces the quantity of waste diverted to treatment facilities and further lowers treatment costs.

Companies invest in wastewater treatment and reuse not just to comply with effluent standards but because product recycling and raw material recovery benefit in terms of reputation. In contrast to agriculture, a small fraction of industrial waters only is actually consumed, and the most part is discharged as wastewater. The ability to reuse water, regardless of whether the intent is to augment water supplies or manage nutrients in treated effluent, has positive benefits that are also the key motivators for implementing reuse programmes. These benefits include:

(a) Improved agricultural production
(b) Reduced energy consumption associated with production, treatment and dis-
    tribution of water
(c) Significant environmental benefits, such as reduced nutrient loads to receiving
    waters due to reuse of the treated effluents [12].

Industrial wastewater treatment has taken place in a series of development
phases starting from direct discharge to recycling and reuse. This development has
been slow when considering the growing awareness of environmental degradation,
public pressure, implementation of increasingly stringent standards and industrial
interest in waste recycling. The declining supply and higher costs of raw water is
also forcing industry to implement recycling technologies. Many industries are now
concentrating on methods to cut down potable water intake and reduce discharge of
polluted effluents. In particular, wastewater reuse has become increasingly important in water resource management for both environmental and economic reasons.

Wastewater reuse has a long history of applications, primarily in agriculture, and
additional application areas, including industrial, household and urban options, are
becoming more prevalent. However, this practice also has its risks and benefits,
which should be critically analysed before taking the decision to either use raw
wastewater directly or use them after treatment. This aspect should be analysed with
reference to local conditions and requirements as wastewater quality and water use
are different in individual countries and regions. Therefore, in order to optimise
water use and cost reduction potential, it is beneficial to analyse both the quality and
the quantity of source effluents against potential reuse applications and water
quality requirements. Appropriate technology and its availability should also be
taken into consideration. Moreover, the control and the continuous improvement of
existing practices have to be taken into account. The level of required treatments for
reclaimed water depends on the intended use [13].

Water reuse applications can be designed for indirect or direct reuse. At present,
reclaimed water is more commonly used for non-drinking purposes, such as agriculture, landscape and park irrigation. Other major applications include greywater
for cooling towers, power plants and oil refineries, toilet flushing, dust control,
construction activities, concrete mixing and artificial lakes.

1.3 Direct Reuse

Direct reuse involves treated wastewaters as potable or process waters: it is a
technically feasible option for agricultural and some industrial purposes (recycled
water within the same industrial process), with or without treatment to meet specific
quality requirements. Some wastewater streams also contain useful materials, such
as organic carbon and nutrients like nitrogen and phosphorous. The use of
nutrient-rich water for agriculture and landscaping may lead to a reduction of
fertiliser applications. Estimations revealed an annual production of 30 million tons

of wastewater in the world, and 70% of this amount is consumed as an agricultural fertiliser and irrigation source [14]. This practice for crop production has gained a certain acceptance worldwide [15] as an economic alternate that could substitute nutrient needs [16, 17] and water requirement of crop plants.

For example, it has been reported that 73,000 ha were irrigated with wastewater during early nineties in India, and presently the area under this irrigation techniques is on the rise [18]. Should the quality be not suitable for direct use, wastewater would necessarily be reclaimed with adequate treatment, or used after dilution with clean water or other higher quality wastewater; indirect use is one of the water recycling applications that has developed, largely as a result of advances in treatment technology.

1.4   Indirect Reuse

Indirect reuse involves the reclamation and treatment of water from wastewater and the eventual returning of it into the natural water cycle (creeks, rivers, lakes and aquifers) or into a receiving body; therefore, this water may be re-treated for use within a plant. The advantage of indirect reuse is essentially the possibility of significant control measure for receiving waters (dilution), provided that contaminant levels in the receiving water are lower than those in the recycled water.

For the above-mentioned reasons, water quality requirements will need to be tailored appropriately. Therefore, the minimisation of risks and the creation/implementation of necessary control measures in place—e.g. water safety plans and Hazard Analysis and Critical Control Point (HACCP) plans—are critical. The purpose of monitoring is to demonstrate that the management system and related treatments are functioning according to design and operating expectations. Expectations should be specified in management systems, according to HACCP or water safety plan approaches.

The Codex Alimentarius framework of risk analysis has been accepted and is recommended as the basis on which this document might be used [19]. The risk analysis process consists of three components: risk assessment, risk management and risk communication. Risk assessment is dependent upon the correct identification of the hazards, the quality of used data and the nature of assumptions made to estimate risk levels. Risk communication should assure the continuous information exchange among all involved parties throughout the entire process, while the monitoring programme has to be written on available regulatory norms and should permit requirements established for the system. This programme not only must address those elements needed to verify the product water, but also must support overall production efficiency and effectiveness.

1.5   Wastewater Reuse Guidelines

The main purpose of reuse criteria is to protect the community and to minimise environmental damages. Reuse guidelines have been issued in the USA, South Africa, Australia, Japan, several Mediterranean basin countries and the European Union. With relation to these documents, the most accepted guidelines appear those published by the United States Environmental Protection Agency [20]. The World Health Organization (WHO) guidelines, issued in 2006, propose a flexible approach of risk assessment and risk management linked to health-based targets that can be established at a realistic level under local conditions. The approach has to be backed up by strict monitoring measures [21].

Wastewater may have both risks from pathogen agents and chemical contaminants from industrial discharges or storm water run-off. WHO guidelines provide maximum tolerable soil concentrations of various toxic chemicals based on human exposure through the food chain. For irrigation water quality, WHO refers to the Food and Agriculture Organization (FAO) guidelines [22]. These guidelines do not specifically address how to reduce chemical contaminants from wastewater for use in irrigation.

Basically, exposure to untreated wastewater is a likely contributor to the burden of diarrhoeal disease worldwide [23]. Epidemiological studies suggest that exposure pathways to the use of wastewater in irrigation can lead to significant infection risk for consumers or populations living near suspect wastewater irrigation sites. These sites may be exposed to aerosols from untreated wastewater and at risk of bacterial and viral infections: several epidemiological investigations have found notable parasitic, diarrhoeal and skin infection risks in farmers and their families living directly in contact with wastewater [21]. Also, excess diarrhoeal diseases and cholera, typhoid and shigellosis outbreaks have been associated with the consumption of wastewater-irrigated vegetables eaten uncooked [21].

1.6   Chemical and Physical Features of Wastewater from Food-Related Activities

In general, the major types of food processing industries associated with high consumption of freshwater are represented by meat-processing plants: the demand of used water is reported to be 24% [24]. On the other hand, the so-called water footprint is variegated in other food and beverage sectors, including the simple crop production: with relation to this sub-group, the higher demand is reported for rice, wheat and maize (21, 12 and 9%, respectively) on the total amount of needed water for crop production worldwide [25].

1.6.1   Slaughterhouses and Related Wastewater

Slaughterhouses Wastewater (SWW) has been considered as an industrial waste in the category of agricultural and food industries and classified as one of the most harmful wastewaters to the environment by the United States Environmental Protection Agency (US EPA).

Slaughterhouses are part of a large industry, which is common to numerous countries worldwide where meat is an important part of their diet. In fact, meat is the first-choice source of animal protein for many people worldwide [25]. The total estimated consumption of meat (chicken, turkey, veal, lamb, beef, pork) in the USA was 101 kg^{-1} capita in the year 2007 [26]. In addition, the consumption of meat is continuously increasing worldwide, particularly in developing countries [27, 28]. The global meat production was doubled in the last three decades, from 2002 to 2007, and the annual global production of beef was increased by 29% over eight years [28]. Furthermore, the production of beef has been increasing continuously in recent years, mostly in India and China, due to income increases and the shift towards a western-like diet rich in proteins [29].

As a result, it can be inferred that the number of slaughterhouse facilities will increase, resulting in a greater volume of high-strength wastewater to be treated. The slaughterhouse industry is the major consumers of freshwater among food and beverage processing facilities [24].

In meat processing, water is used primarily for carcass washing after hide removal from cattle, calves and sheep or hair removal from hogs and again after evisceration, for cleaning, and sanitising of equipment and facilities, and for cooling of mechanical equipment such as compressors and pumps including carcass blood washing, equipment sterilisation and work area clearing. A large water amount is used for different operations such as hog scalding. The rate of water use and wastewater generation can be highly variable often meat-processing facilities work in two different moments: killing and processing shift, followed by cleaning operations.

Elevated consumption of high-quality water, which is an important element of food safety, is often characteristic of the meat-processing industry. SWW composition varies significantly depending on the diverse industrial processes and specific water demand [24, 31]. Abattoir industries produce significant volumes of wastewater due to slaughtering and cleaning of slaughterhouse facilities and meat-processing plants. In particular, the meat-processing industry uses 24% of the total freshwater consumed by the food and beverage industry and up to 29% of that consumed by the agricultural sector worldwide [24, 27, 32].

SWW from the slaughtering process is considered detrimental worldwide due to its complex composition of fats, proteins and fibres [33]. Abattoir wastewater contains high amounts of organic material and consequently high biochemical oxygen demand (BOD) and chemical oxygen demand (COD) values due to the presence of blood, tallow and mucosa.

Meat industry wastewater may also have a high content of nitrogen (from blood) and phosphorus, total suspended solids (TSS) [31, 34]; consequently, SWW discharge may cause deoxygenation of rivers and contamination of groundwater [11]. Further, detergents and disinfectants used for cleaning activities have to be considered because of the presence of pathogenic and non-pathogenic microorganisms and parasite eggs [35]. Meat-processing wastewaters also contain a variety of mineral elements, some of which are present in the water that is used for processing meat. Manure—especially hog manure—may be a significant source of copper, arsenic and zinc, because these constituents are commonly added to hog feed. Due to the presence of manure in meat-processing wastewaters, microbial loads ascribed to total coliforms, faecal coliforms and faecal streptococci are generally found as several million colony forming units per 100 mL. Although members of these groups of microorganisms generally are not pathogenic, they do indicate the possible presence of pathogens of enteric origin such as *Salmonella* spp., *Escherichia coli* O157:H7, *Shigella* spp. and *Campylobacter jejuni*. They also indicate the possible presence of gastrointestinal parasites including *Ascaris* sp., *Giardia lamblia*, *Cryptosporidium parvum* and enteric viruses. In addition to the presence of pathogenic microorganisms, antibiotics used to control pathogens and ensure livestock weight advancement and disease prevention can be found: these substances are released during the evisceration process [36].

1.6.2   Beverage Industries and Related Wastewater

The beverage industry, and important subcategory of the food sector, supplies a range of products from alcoholic (winery, vinasses, molasses and spirits) and brewery to non-alcoholic (fruit juices, vegetable juice, mineral water, sparkling water, flavoured water and soft drinks) beverages [37, 38]. As the global consumption of soft drinks continues to grow (687 billion L in 2013), the global value reaches 830 billion $ [38]. Beverage industry's wastewater could be originated from different individual processes such as bottle washing, product filling, heating or cooling and 'cleaning-in-place' systems, beverage manufacturing, sanitising floors including work cells, cleaning of zones and piping networks [37, 39].

One basic cause of freshwater wastage is the reuse of glass bottles which requires a huge expense of water as rinsing and cleansing agent, before containers are refilled. This treatment has to be necessarily conducted with the aim of removing microorganisms and chemicals to render bottles safe for the human health. Also, different chemicals used for washing of bottles may include sodium hydroxide, detergents and chlorine solution. Washing of bottles is usually done in different stages: pre-rinse, pre-wash, caustic wash and final rinsing. It should be considered that 50% of the total wastewater produced by the beverage industry comes from bottle washing process [37, 40]. In general, critical values for beverage industry wastewater parameters—COD, BOD, TSS, total dissolved solids and total Kjeldahl-determined nitrogen—are normally high [39]. The amount of total nitrogen, total phosphorus and pH can vary depending on used chemicals (nitric acid, phosphoric acid and caustic soda) [41].

1.6.3   Alcoholic Beverages Industries and Related Wastewater

Distilleries, wineries and breweries produce alcoholic beverages. They have strong similarities in terms of manufacturing processes, fermentation and separation operations [42]. As a result, they are high consumers of freshwater and thus produce high volumes of wastewater. The disposal of the untreated wastewater from distillery, winery and brewery industries is considered an environmental hazard worldwide: should the wastewater be discharged into the environment without treatment, salination and eutrophication of freshwater resources would be observed.

1.6.4   Distillery Companies and Related Wastewater

Distillery wastewater refers to wastewater, which is generated from alcohol distilleries. On average, 8–15 L of wastewater is generated for every litre of produced alcohol [45, 46]. Distillery wastewater, generated from the distillation of fermented mash, is dark brown in colour; it contains acidic high organic matter, and with unpleasant odours. The amount of pollution produced from distillery wastewaters depends on the quality of molasses, feedstock, location, characteristics of the distillery manufacturing process and the distillation process that is used to produce ethanol [45]. The BOD/COD ratio of distillery wastewater is considered to be high if >0.6 [42].

1.6.5   Winery Companies and Related Wastewater

Wine production is one of the most important agricultural activities at present [47]. The wine industry can be separated into two sub-categories depending on the specific activity: production of winery wastewater and by-products, and recycling of winery by-products within wine distilleries [42]. Wine production requires a considerable amount of resources such as water, energy, fertilisers and organic amendments; on the other side, it produces a large wastewater amount [48]. This wastewater is one of the final results, in brief, of a number of activities that include: cleaning of tanks, washing of floors and equipment, rinsing of transfer lines, barrel cleaning, off wine and product losses, bottling facilities, filtration units and rainwater diverted into, or captured in the wastewater management system [49].

Each winery is unique with regard to the volume of wastewater generated. In fact, many factors—the working period (i.e. vintage, racking, bottling), the wine making process and the technology applied for red and/or white wine production, etc.—have to be taken into account [50, 51]. In general, it may be affirmed that a winery produces 1.3–1.5 kg of residues per litre of produced wine and 75% of this amount is winery wastewater [52]; on the other side, the generation of winery wastewater is seasonal [42]. Winery effluents are generally biodegradable; BOD/COD ratio tends to be higher during the 'vintage' period because of the presence of molecules such as sugars and ethanol [53].

The environmental impact of wastewater from the wine industry is notable (i.e. pollution of water, eutrophication, degradation of soil and damage to vegetation arising from wastewater disposal practices, odours and air emissions resulting from the management of wastewater), mainly due to the high organic load and large produced volumes [54].

1.6.6   Non-alcoholic Beverages Production and Related Wastewater

The non-alcoholic beverage industry generates wastewater composed of various blends of chemicals[1] [37, 40]. Syrups are reported to be the main pollutants in the non-alcoholic beverage industry wastewater [40] because of the production of pollutants rich in sucrose and often derived from different operations (juice production, cleaning of zones and pipes).

1.6.7   Dairy Industry and Related Wastewater

The dairy industry is one of the main sources of industrial effluent generation in Europe [55]. This sector is based on processing and manufacturing operations of raw milk into products such as yoghurt, ice cream, butter, cheese and various types of desserts by means of different processes, such as pasteurisation, coagulation, filtration, centrifugation, chilling [56]. The dairy industry need for water is huge: in fact, water is used throughout all steps such as sanitisation, heating, cooling milk processing, cleaning, packaging and cleaning of milk tankers.

The dairy industry is subdivided into several sectors associated to the production of contaminated wastewaters. These effluents have different features, depending on

---

[1]These substances include fructose, glutose, sucrose, lactose, artificial sweeteners, fruit juice concentrates, flavouring agents, dissolved carbon dioxide/carbonic acid, bicarbonates, flavourings, colouring additives (caramel and synthetic dye-stuff), preservatives (phosphoric acid and tartaric acid) and mineral salts that are used during production.

final products; the generated volume is quite variable depending on the different types of industry, techniques, processes used in the manufacturing plant and equipment products [57]. Most of the wastewater volume generated in the dairy industry results from clearing of transport lines and equipment between production cycles, cleaning of tank trucks and washing of milk silos [58]. Dairy wastewaters are also characterised by wide fluctuations in flow rates, related to discontinuity in the production cycles of different products [59].

The dairy industry generates huge wastewater amounts, approximately 0.2–10 L of waste per litre of processed milk [60]. In general, the composition of these wastewaters is correlated with high BOD and COD values representing the high organic content[2] [61]. Cheese effluents represent a significant environmental impact in the dairy industry because of their physicochemical features: these effluents exhibit BOD/COD ratio (index of biodegradability) values typically in the range 0.4–0.8 leading to high dissolved oxygen consumption in water bodies.

Lactose and fat contents can be considered as the main responsible for COD and BOD high values. In industrial dairy wastewaters, nitrogen originates mainly from milk proteins, and it is present in various forms: either an organic nitrogen (proteins, urea, nucleic acids) or as ions such as ammonium nitrate and nitrite ions. Phosphorus is found mainly in inorganic forms, as orthophosphate and polyphosphate compounds, as well as organic forms [62, 63].

Waste control is an important aspect of resource management control and an essential part of dairy food plant operations [64]. With their notable concentration of organic matter, these effluents may create serious problems of organic burden on the local municipal sewage treatment systems. Because to the total nitrogen and phosphorus high contents, cheese effluents pose a considerable risk of eutrophication in receiving waters, particularly in lakes and slow-moving rivers [64].

1.6.8 Agro-industrial Wastewater

In the last few years, the need to increase agricultural productivity of the ever-increasing population worldwide has constantly increased. The intensification of agricultural practices leads adverse side effects on critical status of the environment through land usage and soil degradation, water consumption, eutrophication and water pollution, monocultures that cause biodiversity loss and introduction of hazardous chemicals through synthetic pesticides and mineral fertilisers and pesticides. Agriculture is the main user of limited freshwater resources in the world.

On a global scale, 80 ± 10% of all freshwater withdrawals (from lakes, rivers, underground aquifers, etc.) are used in agriculture. More than 40% of the food

---

[2]These values are justified because of the presence of carbohydrates, mainly lactose, as well as less biodegradable proteins, lipids, minerals, high concentrations of suspended solids, suspended oils and grease easily degradable.

production comes from irrigated land; as a result, 70% of freshwaters taken from rivers and groundwater are used for irrigation [65]. It may be forecasted that an additional amount of food products (+60%) will be needed between 2016 and 2050 with the aim of satisfying the demand of an eventual population exceeding nine billion people.

The clear result is that agricultural water use is increasing the severity of water scarcity in some areas and causing water scarcity even in areas that are relatively well endowed with water resources [3, 4]. In this contest, a serious point-source contamination of natural water resources is constituted by fruit/vegetable-packaging plants. Large volumes of effluents and solid waste derive from industrial fresh packing and processing of fruits and vegetables; the demand for water occurs in very specific, relatively short temporal periods. The seasonal nature of processed products can explain the remarkable difference in pollution loads that are eliminated throughout the year.

The special case history is represented by 'Fourth Range' (minimally processed) products. These foods consist of vegetables having been cleaned, are peeled, washed, cut, packed in bags or trays and sold as ready to use fresh foods. The entire treatment and packaging cycle relies on the use of water. The amount of freshwater is huge in relation to the weight of final product. It is in any case a high volume of fresh drinking water, which is at the end of the processing cycle considered wastewater.

Moreover, it has to be considered that wastewaters from the fruit-packaging industry represent an important source of contamination by pesticides. In the absence of effective depuration methods, these fluids are discharged in municipal wastewater treatment plants (alternatively, they can be found in lands) [66]. Pesticides like thiabendazole, imazalil, ortho-phenylphenol or antioxidants such as diphenylamine and ethoxyquin are used to minimise production losses due to fungal infestations or physiological disorders during storage [67, 68].

Postharvest treatments of fruits result in the production of large wastewater volumes which are characterised by low BOD/COD values, but high concentrations of pesticides should advise the preliminary detoxification prior to environmental release [69].

References

1. WHO/UNICEF (2015) Water, sanitation and hygiene—WASH Post (2015) WHO/UNICEF Joint Monitoring Programme (JMP) for water supply and sanitation. World Health Organization (WHO), Geneva, and the United Nations International Children's Emergency Fund (UNICEF), New York
2. Skouteris G, Hermosilla D, López P, Negro C, Blanco A (2012) Anaerobic membrane bioreactors for wastewater treatment: a review. J Chem Eng 198–199:138–148. doi:10.1016/j.cej.2012.05.070
3. UN-Water (2012) Managing water under uncertainty and risk. The United Nations World Water Development Report 4: World Water Assessment Programme (WWAP), Paris, France

4. Valta K, Moustakas K, Sotiropoulos A, Malamis D, Haralambous KJ (2016) Adaptation measures for the food and beverage industry to the impact of climate change on water availability. Desalin Water Treat 57(5):2336–2343. doi:10.1080/19443994.2015.1049407

5. Food Drink Europe (2012) Data & trends of the European food and drink industry 2011. Food Drink Europe, Brussels. Available http://www.fooddrinkeurope.eu/uploads/publications_documents/Final_Data__Trends_30.4.2012.pdf. Accessed 30 Mar 2017

6. Gardner-Outlaw T, Engleman R (1997) Sustaining water, easing scarcity: a second update. Population Action International, Washington, D.C

7. Valta K, Kosanovic T, Malamis D, Moustakas K, Loizidou M (2015) Overview of water usage and wastewater management in the food and beverage industry. Desalin Water Treat 53 (12):3335–3347. doi:10.1080/19443994.2014.934100

8. Hyder Consulting (2013) The Federation House Commitment Progress Report 2013. Waste & Resources Action Programme, Banbury. Available http://www.fdf.org.uk/industry/FHC-ANNUAL-REPORT-2013.pdf. Accessed 30 Mar 2017

9. Coday BD, Xu P, Beaudry EG, Herron J, Lampi K, Hancock NT, Cath TY (2014) The sweet spot of forward osmosis: treatment of produced water, drilling wastewater, and other complex and difficult liquid streams. Desalin 333(1):23–35. doi:10.1016/j.desal.2013.11.014

10. Henze M, van Loosdrecht MCM, Ekama GA, Brdjanovic D (eds) (2008) Biological wastewater treatment. Principles, modelling and design. The International Water Association (IWA) Publishing, London

11. US EPA (2004) Effluent limitations guidelines and new source performance standards for the meat and poultry products point source category. Federal Register 69, 173:54476–54555. United States Environmental Protection Agency (US EPA), Washington, D.C

12. Simons GG, Bastiaanssen WW, Immerzeel WW (2015) Water reuse in river basins with multiple users: a literature review. J Hydrol 522:558–571. doi:10.1016/j.jhydrol.2015.01.016

13. Ong CN (2016) Water reuse, emerging contaminants and public health: state-of-the-art analysis. Int J Water Resour D 32(4):514–525. doi:10.1080/07900627.2015.1096765

14. Cheraghi M, Lorestani B, Yousefi N (2009) Effect of waste water on heavy metal accumulation in Hamedan Province Vegetables. Int J Bot 5:190–193

15. Singh KP, Mohon D, Sinha S (2004) Impact assessment of treated/untreated wastewater toxicants discharge by sewage treatment plant on health, agricultural and environmental quality in waste water disposal area. Chemosph 55(2):227–255. doi:10.1016/j.chemosphere.2003.10.050

16. Campbell CA, Davidson HR (1979) Effect of temperature, nitrogen fertilization and moisture stress on growth, assimilate distribution and moisture use by Manitou spring wheat. Can J Plant Sci 59(3):603–626. doi:10.4141/cjps79-098

17. Rusan MJM, Hinnawi S, Rousan L (2007) Long term effect of waste water irrigation of forage crops on soil and plant quality parameters. Desalin 215:143–152. doi:10.1016/j.desal.2006.10.032

18. Strauss M, Blumenthal UJ (1990) Human waste use in agriculture and aquiculture: utilization practices and health perspectives. IRCWD Report 09/90, p. 48. International Reference Centre for Waste Disposal (IRCWD), Duebendorf

19. Codex Alimentarius Commission (2006) Codex Alimentarius Commission—Procedural Manual, nineteenth edn. Joint FAO/WHO Food Standards Programme, the World Health Organization (WHO), Geneva, and the Food and Agriculture Organization (FAO) Rome

20. US EPA (2012) Guidelines for water reuse. EPA/600/R-12/618. US EPA, Office of Research and Development, Office of Water, Washington, D.C., National Risk Management Research Laboratory, Cincinnati, and United States Agency for International Development, Washington, D.C

21. WHO (2006) Guidelines for the safe use of wastewater, excreta and greywater. World Health Organization WHO, Geneva

22. Steduto P, Hsiao TC, Fereres E, Raes D (2012) Crop yield response to water. FAO Irrigation and Drainage paper 66. FAO, Rome

23. WHO (2004) The global burden of disease report: 2004 update. World Health Organization WHO, Geneva. Available http://www.who.int/healthinfo/global_burden_disease/2004_report_update/en/index.html. Accessed 30 Mar 2017
24. Bustillo-Lecompte CF, Mehrvar M (2015) Slaughterhouse wastewater characteristics, treatment, and management in the meat processing industry: a review on trends and advances. J Environ Manag 161:287–302. doi:10.1016/j.jenvman.2015.07.008
25. Hoekstra AY, Chapagain AK (2006) Water footprints of nations: water use by people as a function of their consumption pattern. In: Craswell E, Bonnell M, Bossio D, Demuth S, Van De Giesen N (eds) Integrated assessment of water resources and global change, Springer, Netherlands. pp 35–48. doi:10.1007/978-1-4020-5591-1_3
26. Dave D, Ghaly AE (2011) Meat spoilage mechanisms and preservation techniques: a critical review. Am J Agric Econ Biol Sci 6(4):486–510
27. Mekonnen MM, Hoekstra AY (2012) A global assessment of the water footprint of farm animal products. Ecosyst 15(3):401–415. doi:10.1007/s10021-011-9517-8
28. FAO (2013) Food outlook—biannual report on global food markets. FAO Trade and Market Division, Rome. Available, Italy. http://www.fao.org/docrep/019/i3473e/i3473e.pdf. Accessed 30 Mar 2017
29. Pingali P (2007) Westernization of Asian diets and the transformation of food systems: implications for research and policy. Food Policy 32(3):281–298. doi:10.1016/j.foodpol.2006.08.001
30. Wang LK, Hung YT, Lo HH, Yapijakis C (eds) (2006) Waste treatment in the food processing industry. CRC Press, Boca Raton
31. Bustillo-Lecompte CF, Mehrvar M, Quiñones-Bolaños E (2014) Cost-effectiveness analysis of TOC removal from slaughterhouse wastewater using combined anaerobic–aerobic and UV/H₂O₂ processes. J Environ Manag 134:145–152. doi:10.1016/j.jenvman.2013.12.035
32. Gerbens-Leenes PW, Mekonnen MM, Hoekstra AY (2013) The water footprint of poultry, pork and beef: a comparative study in different countries and production systems. Water Resour Ind 1(2):25–36. doi:10.1016/j.wri.2013.03.001
33. Wu PF, Mittal GS (2011) Characterization of provincially inspected slaughterhouse wastewater in Ontario, Canada. Can Biosyst Eng 54(6):9–18
34. Barrera M, Mehrvar M, Gilbride KA et al (2012) Photolytic treatment of organic constituents and bacterial pathogens in secondary effluent of synthetic slaughterhouse wastewater. Chem Eng Res Des 90(9):1335–1350. doi:10.1016/j.cherd.2011.11.018
35. Mass DI, Masse L (2000) Characterization of wastewater from hog slaughterhouses in Eastern Canada and evaluation of their in-plant wastewater treatment systems. Can Agric Eng 42 (3):139–146
36. Us EPA (2002) Development document for the proposed effluent limitations guidelines and standards for the meat and poultry products industry point source category (40 CFR 432). US EPA, Office of Water, Washington, D.C
37. Haroon H, Waseem A, Mahmood Q (2013) Treatment and reuse of wastewater from beverage industry. J Chem Soc Pak 35(1):5–10
38. Euromonitor (2014) Passport: global market information database. Euromonitor International. http://www.euromonitor.com
39. Agana BA, Reeve D, Orbell JD (2013) Performance optimization of a 5 nm TiO₂ ceramic membrane with respect to beverage production wastewater. Desalin 311:162–172. doi:10.1016/j.desal.2012.11.027
40. Hsine EA, Benhammou A, Pons MN (2005) Water resources management in soft drink industry-water use and wastewater generation. Environ Technol 26(12):1309–1316. doi:10.1080/09593332608618605
41. Nyilimbabazi N, Banadda N, Nhapi I (2011) Characterization of brewery wastewater for reuse in Kigali, Rwanda. Open Environ Eng J 4:89–96
42. Fillaudeau L, Blanpain-Avet P, Daufin G (2006) Water, wastewater and waste management in brewing industries. J Clean Prod 14(5):463–471. doi:10.1016/j.jclepro.2005.01.002

43. Simate GS, Cluett J, Iyuke SE Musapatika ET, Ndlovu S, Walubita LF, Alvarez AE (2011) The treatment of brewery wastewater for reuse: state of the art. Desalin 273(2):235–247. doi:10.1016/j.desal.2011.02.035

44. Kanai M, Ferre V, Wakahara S, Yamamoto T, Moro M (2010) A novel combination of methane fermentation and MBR-Kubota submerged anaerobic membrane bioreactor process. Desalin 250(3):964–967. doi:10.1016/j.desal.2009.09.082

45. Mohana S, Acharya BK, Madamwar D (2009) Distillery spent wash: treatment technologies and potential applications. J Hazard Mater 163(1):12–25. doi:10.1016/j.jhazmat.2008.06.079

46. Satyawali Y, Balakrishnan M (2008) Treatment of distillery effluent in a membrane bioreactor (MBR) equipped with mesh filter. Sep Purif Technol 63(2):278–286. doi:10.1016/j.seppur.2008.05.008

47. Hussain M, Cholette S, Castaldi RM (2008) An analysis of globalization forces in the wine industry: implications and recommendations for wineries. J Glob Market 21:33–47. doi:10.1300/J042v21n01_04

48. Arcese G, Lucchetti MC, Martucci O (2012) Analysis of sustainability based on life cycle assessment: an empirical study of wine production. J Environ Sci Health B 1(5):1682–1689

49. Ruggieri E, Cadena J, Martinez-Blanco CM, Gasol CM, Rieradevall J, Gabarrell X, Gea T, Sort X, Sánchez A (2009) Recovery of organic wastes in the Spanish wine industry. Technical, economic and environmental analyses of the composting process. J Clean Prod 17 (9):830–838. doi:10.1016/j.jclepro.2008.12.005

50. Bustamante MA, Moral R, Paredes C, Pérez-Espinosa A, Moreno-Caselles J, Pérez-Murcia MD (2008) Agrochemical characterisation of the solid by-products and residues from the winery and distillery industry. Waste Manag 28(2):372–380. doi:10.1016/j.wasman.2007.01.013

51. Devesa-Rey R, Vecino X, Varela-Alende JL, Barral MT, Cruz JM, Moldes AB (2011) Valorization of winery waste vs. the costs of not recycling. Waste Manag 31(11):2327–2335. doi:10.1016/j.wasman.2011.06.001

52. Lucas MS, Peres JA, Li Puma G (2010) Treatment of winery wastewater by ozone-based advanced oxidation processes (O_3, O_3/UV and $O_3/UV/H_2O_2$) in a pilot-scale bubble column reactor and process economics. Sep Purif Technol 72(3):235–241. doi:10.1016/j.seppur.2010.01.016

53. Mosteo R, Sarasa J, Ormad MP, Ovelleiro JL (2008) Sequential solar photo-Fenton-biological system for the treatment of winery wastewaters. J Agric Food Chem 56(16):7333–7338. doi:10.1021/jf8005678

54. Chatzilazarou A, Katsoyannos E, Gortzi O, Lalas S, Paraskevopoulos Y, Dourtoglou E, Tsaknis J (2010) Removal of polyphenols from wine sludge using cloud point extraction. J Air Waste Manag Assoc 60(4):454–459. doi:10.3155/1047-3289.60.4.454

55. Demirel B, Yenigun O, Onay TT (2005) Anaerobic treatment of dairy wastewaters: a review. Process Biochem 40(8):2583–2595. doi:10.1016/j.procbio.2004.12.015

56. Rivas J, Prazeres AR, Carvalho F, Beltrán F (2010) Treatment of cheese whey wastewater: combined coagulation-flocculation and aerobic biodegradation. J Agric Food Chem 58 (13):7871–7877. doi:10.1021/jf100602j

57. Carvalho F, Prazeres AR, Rivas J (2013) Cheese whey wastewater: characterization and treatment. Sci Total Environ 445–446:385–396. doi:10.1016/j.scitotenv.2012.12.038

58. Kasapgil B, Anderson GK, Ince O (1994) An investigation into the pretreatment of dairy wastewater prior to aerobic biological treatment. Water Sci Technol 29(9):205–212

59. Britz TJ, Van Der Merwe M, Riedel KHJ (1992) Influence of phenol additions on the efficiency of an anaerobic hybrid digester treating landfill leachate. Biotechnol Lett 14 (4):323–328. doi:10.1007/BF01022332

60. Balannec B, Vourch M, Rabiller-Baudry M, Chaufer B (2005) Comparative study of different nanofiltration and reverse osmosis membranes for dairy effluent treatment by dead-end filtration. Sep Purif Technol 42(2):195–200. doi:10.1016/j.seppur.2004.07.013

61. Mukhopadhyay R, Talukdar D, Chatterjee BP, Guha AK (2003) Whey processing with chitosan and isolation of lactose. Process Biochem 39(3):381–385. doi:10.1016/S0032-9592 (03)00126-2
62. Guillen-Jimenez E, Alvarez-Mateos P, Romero-Guzman F, Pereda-Marin J (2000) Bio-mineralization of organic matter as affected by pH. The evolution of ammonium and phosphates. Water Res 34(4):1215–1224. doi:10.1016/S0043-1354(99)00242-0
63. Prazeres AR, Carvalho F, Rivas J (2012) Cheese whey management: a review. J Environ Manag 110:48–68. doi:10.1016/j.jenvman.2012.05.018
64. Arumugam A, Sabarethinam PL (2008) Performance of a three-phase fluidized bed reactor with different support particles in treatment of dairy wastewater. ARPN J Eng Appl Sci 3 (5):42–44
65. Jensen CR, Ørum JE, Pedersen SM, Andersen MN, Plauborg F, Liu F, Jacobsen SE (2014) A short overview of measures for securing water resources for irrigated crop production. J Agron Crop Sci 200(5):333–343. doi:10.1111/jac.12067
66. Kroyer GT (1995) Impact of food processing on the environment—an overview. LWT Food Sci Technol 28(6):547–552. doi:10.1016/0023-6438(95)90000-4
67. Smilanick JL, Mansour MF, Gabler FM, Sorenson D (2008) Control of citrus postharvest green mold and sour rot by potassium sorbate combined with heat and fungicides. Postharvest Biol Technol 47(2):226–238. doi:10.1016/j.postharvbio.2007.06.020
68. Jonkers N, Sousa A, Galante-Oliveira S, Barroso CM, Kohler HPE, Giger W (2010) Occurrence and sources of selected phenolic endocrine disruptors in Ria de Aveiro, Portugal. Environ Sci Pollut Res 17(4):834–843. doi:10.1007/s11356-009-0275-5
69. Santiago DE, Pulido Melián E, Fernández Rodríguez C, Ortega Méndez JA, O. Pérez-Báez SO, Doña-Rodríguez JM (2011) Degradation and detoxification of banana postharvest treatment water using advanced oxidation techniques. Green Sustain Chem 1(3):7067,8. doi:10.4236/gsc.2011.13008

Chapter 2
Wastewater Treatments for the Food Industry: Physical–Chemical Systems

Abstract This chapter provides a general overview of physical–chemical wastewater remediation systems in the food industry. Water reuse systems are becoming more and more interesting and promising technologies, depending on merely quantitative estimations, physical and chemical features of pollutants and the variability of these characteristics, week after week. Different systems are available for the food industry, depending on the final destination or water effluents and peculiar chemical–physical and biological features of the fluids before treatment. Several of these remediation systems can be subdivided into different groups, depending on the desired amount of gross removed matters, or into four categories depending on the peculiar removal operation (physical, chemical, thermal or biological procedures). This chapter is dedicated to the description of physical–chemical wastewater remediation systems only. Biological procedures are not considered here, while physical–chemical techniques are discussed with the possibility of 'hybrid' solutions including biological treatments, if applicable.

Keywords Centrifugation · Evaporation · Filtration · Membrane technology · Remediation · Separation · Wastewater

Abbreviations

BOD Biochemical oxygen demand
COD Chemical oxygen demand
FAO Food and Agriculture Organization of the United Nations

2.1 Introduction to Chemical Wastewater Remediation in the Food Industry. Objectives and Conditions

At present, it may be admitted that water sources are the main concern in several economic areas. Surely, the truthfulness of this affirmation can be observed when speaking of water supplies for food/beverage production and packaging lines.

© The Author(s) 2018   17
M. Barbera and G. Gurnari, *Wastewater Treatment and Reuse in the Food Industry*,
Chemistry of Foods, https://doi.org/10.1007/978-3-319-68442-0_2

For this reason at least, water reuse systems are becoming more and more interesting and promising technologies: generally, discharged water from processing plants can be reused by means of innovative and advanced treatments. However, the final goal can be obtained by means of different strategies, depending on merely quantitative estimations (volumes of wastewaters), chemical features of pollutants (oils, etc.), physical–chemical parameters (biological oxygen demand, solid or liquid pollutants, etc.) and the variability of these characteristics, week after week. On these bases, different systems can be now available for the food industry. Anyway, the right strategy has to be decided on the basis of chemical and biological tests carried out on initial wastewater; the final use of waters is also crucial. Moreover, different chemical systems can be used when speaking of wastewater from food industries for subsequent non-food reuses. Because of the presence of different classes of resistant pollutants, many treatments require often a preliminary adsorption stage.

Actually, the discussion about water reuse systems should take into account a peculiar distinction between technologies designed for the reduction of wastewater and methods/procedures able to reduce the contamination level of existing wastewaters. This distinction has to be taken into account as a preliminary concept or operative definition for wastewater-related treatments [1].

The first category involves preventive measures against the augment of existing wastewaters. Interestingly, these systems are relatively inexpensive (if compared with other treatments) and can easily be put in place in virtually all possible food/beverage plants without size limitations [1]. Our attention is focused on the second group of treatments, also named 'wastewater remediation' systems. However, it should be considered that these treatments may be further subdivided in two different categories depending on the peculiar liquid which should be treated. In fact, waters in the food and beverage industries can be either reused in different sections or subsections of the same plant (before of the final exit to the external sewage) and eliminated as wastewater (this water is directed to publicly owned treatment works) [1]. For this basic reason, the destination of wastewaters defines the best treatment, depending also on the peculiar chemical–physical and biological features of the fluids before treatment.

By a general viewpoint, food wastewaters are the best type of contaminated water when speaking of industrial activities because of the low amount of toxic compounds normally related to the industry of metals or intermediate chemicals (petroleum, plastics, etc.) [1, 2]. However, these fluids have their 'problems' because of their high levels of selected contaminants (minerals, ammonia salts, fats, oils, sugars, starch, etc.). Because of their notable amount of organic matters, wastewaters are also classified on the basis of two different indexes: chemical oxygen demand (COD) and biochemical oxygen demand (BOD). These parameters can give an approximate but correct idea of the state of wastewaters in terms of general contamination. Consequently, input data for wastewaters are often expressed as BOD and COD values, and the same thing is true for output generated data (in terms of BOD and COD values for 'remediated' waters before treatment). The choice of the best remediation treatment should take into account COD and BOD values for the entering wastewater, the level of desired removal (in terms of gross removed matters), plant costs and the desired level of BOD and COD values (with pH, close to neutrality, and analytical results for minerals and other analytes) for exiting waters [1]. With relation to the desired amount of gross removed matters, there is a simple classification which subdivides all processes in three basic categories:

(1) Primary processes. These systems are basically the separation of suspended solids from wastewaters. The aim should be an effluent with notable organic matters and remarkable BOD values
(2) Secondary treatments. These procedures aim to reduce organic loads and the remaining suspended materials in wastewaters from primary processes. As a result, average BOD values should be relatively low for effluents (not more than 30 mg/l) and similar values should be obtained when speaking of suspended solids. In general, secondary processes are based on the biological activity and degradation of pollutants
(3) Tertiary processes. These systems, also defined 'advanced treatments', aim to enhance the chemical and biological quality of effluents to a high-standard values. In other terms, the objective is to obtain water effluents with BOD values and other parameters very low if compared with waters obtained by secondary processes only.

Anyhow, the easier classification of remediation techniques may be offered when speaking of the meaning of the peculiar removal operation. Consequently, remediation systems may be subdivided in [3, 4]:

(a) Physical removal (e.g. filtration). This is a primary process
(b) Chemical systems (e.g. oxidation, coagulation)
(c) Thermal procedures (e.g. oxidation, drying)
(d) Biological removal (e.g. biomass fermentation). Basically, these systems are 'secondary treatments'.

The main difference between biological systems and the other strategies (with the exclusion of separation/concentration, other technologies are substantially 'advanced' or 'tertiary' systems), is based on the degradation of contaminants by microorganisms in the first situation. Soluble and non-soluble pollutants and nutrients based on nitrogen and/or phosphorus are biologically degraded and converted in different and less hazardous compounds.

This chapter is dedicated to the description of physical–chemical wastewater remediation systems only. For this reason, biological procedures are not considered here, while physical–chemical techniques are discussed with the possibility of 'hybrid' solutions including biological treatments, if applicable.

2.2 Physical–Chemical Remediation Systems

The main technologies of wastewater remediation chemical systems in the food industry are generally listed as follows. The subsequent sections give a brief description of each system.

2.2.1 Gravity Separation or Concentration

Basically, these techniques aim to separate solid and semi-solid compounds and materials from wastewaters [1]. In general, this process can be performed by means of bar screens (screening system) and/or with the use of sedimentation basins (mechanical processes) [4]. The key is the density of pollutants (oils and grease are lighter than water; suspended solids are agglomerated on the bottom of basins). Chemical treatments may be also used. Anyway, the aim is to eliminate 50% or more of the total suspended solids [5], more than 60% of oils and grease, with the consequent reduction of BOD values after five testing days (also named BOD_5). Unfortunately, this treatment cannot remove colloidal and dissolved compounds, while nitrogen- and phosphorus-associated organic molecules and heavy metals can be notably reduced [5]. It should be noted that certain primary processes can be coupled with chemical and also biological treatments: in the last situation, the obtained sludge is digested anaerobically with methane production and recovery. Other solutions are also possible [5].

2.2.2 Evaporation

This process is a typical concentration process: it can be considered when speaking of wastewaters with inorganic salts in notable amounts. Basically, salts and other compounds including heavy metals are concentrated and recovered for other uses while distilled water is normally obtained with very good chemical and physical features, and consequently reusable. Different evaporation machines (mechanical equipment, evaporator ponds, with vertical or horizontal geometry or with forced circulation of fluids) can be used; costs may be high depending on the amount of treated fluids [1]. Anyway, these systems may require some maintenance and additional treatment systems because of possible defects (fouling is only one of possible examples).

2.2.3 Centrifugation

This separation process is useful when speaking of wastewaters with notable oil amounts and/or particle sizes under 5000 μm [1]. Basically, it is only a simple centrifugation process with different machines and costs (depending on the geometry and the amount of treated fluids). It has to be noted that wastewaters with particle sizes exceeding 5000 μm may be treated with this system on condition that high-sized bodies and compounds are previously separated.

In the sector of 'Fourth Range' (minimally processed) products various methods of water treatment systems are used in order to obtain less pollution: machinery is dealing with automatic spin dryer (such as cyclones), blowing washed and cascade washing systems.

2.2.4 Filtration and Flotation

Filtration, considered as a pre-treatment or a 'tertiary' process in certain situations, is performed by means of different filters (e.g. cartridges, membrane systems, generally used as a pre-treatment step or a final wastewater 'polishing' step before discharge. Several different types of filters exist, including granular-media, cartridge and pre-coated filters with diatomaceous earth). This system, very used in the food industry (e.g. filtration of brine solutions for cheeses), is useful on condition that particle sizes are >1 μm [1].

In some processes, such as in the production of Fourth Range foods, the intermediate processing system based on the utilisation of clean water could be treated following natural methods, like a filtration on natural sand beds. Using the equations of Darcy the flow speed and related time of transit of the fluids can be calculated in order to obtain water free from contaminants.

Another separation treatment uses the adherence of oils and grease to gas bubbles when pumped in wastewaters (dissolved or induced air flotation systems); superficial agglomerations may be eliminated by skimming [1].

2.2.5 Membrane Technologies

Actually, these systems are an emerging technology in the broader ambit of filtration treatments. Different membranes (materials: polymeric compounds such as polyamides, polycarbonates.) are used depending on the size of pollutants. In general, microfiltration is recommended if particle sizes are <10 mm (target compounds. colloidal compounds, microbial agglomerations). Ultrafiltration systems are recommended when particle sizes are <100 nm (the process is a simple diffusion method): obtained effluents are recovered, while concentrated substances are removed or incinerated [1]. Target molecules are usually colloids, proteins and different emulsions.

Should pollutant sizes be <10 nm, nanofiltration would be recommended. Naturally, this situation is expensive: with relation to wastewaters, the removal of antibiotic substances or the demineralisation of treated waters could be suggested. Finally, reverse osmosis is recommended only when particle sizes are <1 nm, and electrodialysis is used if pure water has to be obtained [1]. Naturally, exigencies of food and beverage industries cannot contemplate all these solutions.

References

1. Anonymous (1997) Wastewater reduction and recycling in food processing operations. State of the Art Report— Food Manufacturing Coalition for Innovation and Technology Transfer. R. J. Philips & Associates, Inc., Great Falls
2. Nini D, Gimenez-Mitsotakis P (1994) Creative solutions for bakery waste effluent. American Institute of Chemical Engineers. Symposium Series 300, 90:95–105
3. Henze M, van Loosdrecht MCM, Ekama GA, Brdjanovic D (eds) (2008) Biological wastewater treatment. Principles, modelling and design. The International Water Association (IWA) Publishing, London
4. Munter R (2000) Industrial wastewater treatment. In: Lundin LC (ed) Water use and management. Uppsala University, Uppsala
5. Pescod MD (1992) Wastewater treatment and use in agriculture—FAO Irrigation and Drainage Paper 95. Food and Agriculture Organization of the United Nations (FAO), Rome. Available http://www.fao.org/docrep/t0551e/t0551e00.htm. Accessed 29 Mar 2017

Chapter 3
Wastewater Treatments for the Food Industry: Biological Systems

Abstract This chapter provides a general overview of biological wastewater remediation systems in the food industry. Water reuse systems are becoming more and more interesting and promising technologies, depending on merely quantitative estimations, physical and chemical features of pollutants and the variability of these characteristics, week after week. Different systems are available for the food industry. Several of these remediation systems may be subdivided into four categories depending on the peculiar removal operation, including biological systems. Biological techniques aim to reduce organic loads and the remaining suspended materials in wastewaters from primary processes (after a preliminary removal of oils and solids) by means of aerobic, anaerobic or hybrid solutions. Soluble and non-soluble pollutants and nutrients based on nitrogen and/or phosphorus are biologically degraded and converted in different and less hazardous compounds.

Keywords Aerobic metabolism · Anaerobic metabolism · BOD · COD · Oxidation · Separation · Wastewater

Abbreviations

BOD   Biochemical oxygen demand
COD   Chemical oxygen demand

3.1 Introduction to Wastewater Bioremediation in the Food Industry: Objectives and Conditions

Water sources are the main concern in several economic areas. Surely, the truthfulness of this affirmation can be observed when speaking of water supplies for food/beverage production and packaging lines. For this reason at least, water reuse systems are becoming more and more interesting and promising technologies: generally, discharged water from processing plants can be reused by means of innovative and advanced treatments. However, the final goal can be obtained by

© The Author(s) 2018

M. Barbera and G. Gurnari, *Wastewater Treatment and Reuse in the Food Industry*, Chemistry of Foods, https://doi.org/10.1007/978-3-319-68442-0_3

means of different strategies, depending on merely quantitative estimations (volumes of wastewaters), chemical features of pollutants, physical–chemical parameters and the variability of these characteristics, week after week. On these bases, different systems can be now available for the food industry. Anyway, the right strategy has to be decided on the basis of chemical and biological tests carried out on initial wastewater; the final use of waters is also crucial. Moreover, different chemical systems can be used when speaking of wastewater from food industries for subsequent non-food reuses. Because of the presence of different classes of resistant pollutants, many treatments require often a preliminary adsorption stage.

A preliminary distinction between technologies designed for the reduction of wastewater and methods/procedures able to reduce the contamination level of existing wastewaters has been mentioned briefly in Chap. 2. This distinction has to be taken into account as a preliminary concept or operative definition for wastewater-related treatments [1]. The first category involves preventive measures against the augment of existing wastewaters. Interestingly, these treatments can easily be put in place in virtually all possible food/beverage plants without size limitations [1]. The second category, 'wastewater remediation' system, may be further subdivided into two different categories depending on the treated fluid and the final destination of effluents. The destination of wastewaters defines the best treatment, depending also on the peculiar chemical–physical and biological features of the fluids before treatment.

By a general viewpoint, food wastewaters are classified on the basis of two different indexes: chemical oxygen demand (COD) and biochemical oxygen demand (BOD). Consequently, input data for wastewaters are often expressed as BOD and COD values, and the same thing is true for output generated data (in terms of BOD and COD values for 'remediated' waters before treatment). The choice of the best remediation treatment should take into account COD and BOD values for the entering wastewater, the level of desired removal, plant costs and the desired level of BOD and COD values for exiting waters [1].

In addition, there is a simple classification which subdivides all processes in three basic categories depending on the desired amount of gross removed matters (Sects. 2.1):

(1) Primary processes. These systems are basically the separation of suspended solids from wastewaters. The aim should be an effluent with notable organic matters and remarkable BOD values
(2) Secondary treatments. These procedures aim to reduce organic loads and the remaining suspended materials in wastewaters from primary processes. As a result, average BOD values should be relatively low for effluents (not more than 30 mg/l) and similar performances should be obtained when speaking of suspended solids. In general, secondary processes are based on the biological activity and degradation of pollutants
(3) Tertiary or advanced treatments (for high-standard effluent waters).

Bioremediation systems are substantially secondary processes [2, 3]: the main difference between biological systems and other strategies is based on the degradation of contaminants by microorganisms in the first situation. Soluble and non-soluble pollutants and nutrients based on nitrogen and/or phosphorus are biologically degraded and converted in different and less hazardous compounds.

This chapter is dedicated to the brief description of bio-based wastewater remediation systems only. For this reason, physical–chemical and mechanical techniques are not considered here.

3.2 Preliminary Removal of Oils and Solids

Basically, biological remediation of wastewaters involves the use of bioreactors containing a specific group of active life forms. These active microorganisms may be suspended in the culture media or attached to physical supports [1]. Anyway, the result of biological activity on wastewaters entering bioreactors is the production of carbon dioxide and other catabolites from pollutants bioconversion via aerobic or anaerobic metabolism.

However, the correct performance of such a process (or processes) depends on the 'quality' of entering wastewaters. Consequently, the 'lighter' the fluid into the bioreactor, the better the quality of effluents (in terms of BOD values, pH, absence of specific pollutants, with the exclusion of living microorganisms which should be eliminated in a subsequent step). For this reason, a preliminary or 'primary' step is required with the aim of removing too viscous or rheologically incompatible matters if compared with waters: these materials can really disturb the process [4]. In general, these matters are oils and grease [5]: as an example, a good technical treatment is represented by sedimentation or filtration systems (Sects. 2.2.1 and 2.2.4), although other separation treatments are possible. The aim is to eliminate 50% or more of the total suspended solids and, more than 60% of oils and grease, with the consequent reduction of BOD values after five testing days. Unfortunately, this treatment cannot remove colloidal and dissolved compounds, while nitrogen- and phosphorus-associated organic molecules and heavy metals can be notably reduced.

After the primary treatment, entering wastewaters have to be turned into a good-quality water mass after biological remediation. Basically, aerobic and anaerobic microorganisms are used in this step and related processes are named in the same way, although 'hybrid' processes can be also used and with notable results in certain situations.

3.3   Aerobic Treatments

These systems are well known because of their efficiency and related results: the final products of aerobic activity correspond to inorganic molecules, basically carbon dioxide and water, with the concomitant augment of living microorganisms [5].

This simplified description does not express all possible variation of the general aerobic treatment. Several variables have to be considered, including:

(a) The different supply technology of oxygen
(b) The different rapidity of aerobic metabolism (in other terms, the rapidity of microbial spreading into bioreactors)
(c) The dimension of bioreactors and the concomitant (inversely proportional) amount of active aerobic microorganisms. High-rate processes are carried out into 'little' reactors, but with a notable amount of active life forms.

On these bases, three different aerobic treatments are known and used at present [5]:

(1) The so-called activated sludge technique. This discontinuous system is basically performed by means of a container (the bioreactor) with the wastewater and microorganisms in suspension and a continuous supply of gaseous oxygen (aeration devices are present). As a result, effluents are of acceptable quality, while microorganisms are removed by sedimentation and partially recycled for another process
(2) Biofiltering systems. This approach is based on the use of peculiar surfaces (stones, plastic materials, wood) impregnated with living microorganisms. The resulting active biofilm can 'work' on the wastewater which is continuously or discontinuously supplied (air is also needed). Exiting waters are clarified two times, and the second clarification should give a good-quality water (a part of this effluent is also recycled in the process)
(3) Rotating biological contactors. This technique is based on the same concept of biofilters, but supports for biofilms are rotating discs with a slow speed rate.

With relation to predictable results, biofilm-based systems give better performances than activated sludge. Unfortunately, these systems cannot be able to remove nitrogen, phosphorus-based molecules and non-biodegradable compounds. In general, 70 ± 30 mg per litre of residual chemicals remains (they can be pesticides, artificial chemicals, normal biological catabolites, etc.) [3]. For these reason, the synergy with non-biological techniques is always recommended [5]. Alternatively, more drastic chemical treatments are needed: one of the most known examples in this ambit is represented by oxidation systems [3].

At the end of the process, the mass of living microorganisms has to be roughly removed from the effluent by means of simple sedimentation techniques. This mass generally named biological sludge may be reused by addition to the mass obtained before secondary treatment (at the preliminary stage) and subsequent processing, if required [5].

3.4   Anaerobic Treatments

In contrast with aerobic digestion, anaerobic treatments are substantially based on the biodecomposition of organic pollutants in absence of oxygen with production of the so-called biogas (methane and carbon dioxide). In this ambit, three peculiar bacteria types have been recognised to be useful when speaking of anaerobic digestion of wastewaters [6]:

(a) Fermenting microorganisms. The involved fermentative pathways produce simple organic molecules such as alcohols, carbon dioxide and ammonia
(b) Acetic acid bacteria. These gram-negative life forms can turn carbohydrates or ethyl alcohol into acetic acid (by-products are molecular hydrogen and carbon dioxide)
(c) Methane-producing microorganisms. These extremely important bacteria turn molecular hydrogen and carbon dioxide into methane. Alternatively, acetate ion may be metabolised instead of carbon dioxide.

Consequently, the anaerobic treatment should be seen as a three-stage process: the third stage produces the final reduction product (methane) in high amounts, while the first steps need to be performed because of the evident lack of carbon dioxide (or acetate) and molecular hydrogen. Because of the complexity of involved processes, different solutions exist at present, including high-rate reactors with the following technologies: fluidised bed, anaerobic filter and up-flow anaerobic sludge blanket processes [6].

The reason of the success of anaerobic treatments in comparison with aerobic systems is apparently related to the notable conversion performances of insoluble pollutants at high temperatures and concentrations. Aerobic systems are generally recommended when speaking of soluble pollutants. On the other side, it should be noted that anaerobic life forms need more time than aerobic microorganisms when speaking of conversion speed (the low rapidity of growth is correlated with low speed). Consequently, anaerobic treatment needs to allow a long contact time between target pollutants and involved microorganisms. In addition, attached bioactive microorganisms seem to work better than other solutions, when speaking of anaerobic life forms [6].

3.5   Hybrid Solutions

Because of the different results obtained with aerobic and anaerobic systems and the necessity of ancillary treatments, the 'pure' biological treatment does not exist as a complete remediation process. For these reasons, different chemical systems may be coupled with biological strategies.

As a single example, one of these solutions for 'difficult' wastewaters is the coupling of conventional biological digestion with advanced oxidation techniques and other chemical systems such as chlorination [7, 8]. Interestingly, the oxidation of complex compounds with the production or a more biodegradable mass of pollutants can be applied before biological systems (as a pre-treatment) with good results. On the other hand, advanced chemical oxidation processes—ozonation, Fenton-assisted or photo-assisted membrane processes, etc.—could be performed after biological treatments with the aim of destroying persistent compounds (tertiary processes) in different ambits, including food industries. However, more research is needed at present because of the lack of sufficient information concerning toxicology and biodegradability aspects in coupled strategies; also, economic efficiency should be carefully evaluated [7]. Otherwise, the management of water effluents could be a real problem when speaking of reuse in food industries at least. In fact, food products are already forced to suffer irreversible changes according to the Parisi's Law of Food Degradation [9], and the introduction of potentially contaminated waters in the processing cycle could complicate the production of safe and durable foods.

References

1. Anonymous (1997) Wastewater reduction and recycling in food processing operations. State of the art report—Food manufacturing coalition for innovation and technology transfer. R. J. Philips & Associates, Inc., Great Falls
2. Henze M, van Loosdrecht MCM, Ekama GA, Brdjanovic D (eds) (2008) Biological wastewater treatment. Principles, modelling and design. The International Water Association (IWA) Publishing, London
3. Munter R (2000) Industrial wastewater treatment. In: Lundin LC (ed) Water use and management. Uppsala University, Uppsala
4. Hogetsu A, Ogino Y, Takemika T (2003) Technology transfer manual of industrial wastewater treatment, p. 107–110. Overseas Environmental Cooperation Center, Japan, Ministry of the Environment, Tokyo. Available https://www.env.go.jp/earth/coop/coop/document/male2_e/007.pdf. Accessed 29 Mar 2017
5. Pescod MD (1992) Wastewater treatment and use in agriculture—FAO Irrigation and Drainage Paper 95. Food and agriculture organization of the United Nations (FAO), Rome. Available http://www.fao.org/docrep/t0551e/t0551e00.htm. Accessed 29 Mar 2017
6. Marchaim U (1992) Biogas processes for sustainable development—FAO Irrigation and Drainage Paper 47. Food and agriculture organization of the United Nations (FAO), Rome. Available http://www.fao.org/3/a-t0541e/index.html. Accessed 29 Mar 2017
7. Oller I, Malato S, Sánchez-Pérez J (2011) Combination of advanced oxidation processes and biological treatments for wastewater decontamination—a review. Sci Total Environ 409 (20):4141–4166. doi:10.1016/j.scitotenv.2010.08.061
8. Marco A, Esplugas S, Saum G (1997) How and why combine chemical and biological processes for wastewater treatment. Water Sci Technol 35(4):321–327. doi:10.1016/S0273-1223(97)00041-3
9. Parisi S (2002) I fondamenti del calcolo della data di scadenza degli alimenti: principi ed applicazioni. Ind Aliment 41, 417: 905–919

Chapter 4
Quality Standards for Recycled Water: *Opuntia ficus-indica* as Sorbent Material

Abstract In recent years, increased industrial and agricultural activities and the correlated population growth led to overexploitation of natural resources and the increased generation of various types of pollutants. For these reasons, the hazardous pollution of wastewater is one of the most important environmental problems worldwide. A wide range of wastewater treatment technologies are available; however, some disadvantages are often reported. Hence, there is a constant need to search for an efficient, low-cost and alternative wastewater treatment. Recently, several biosolids have been considered for pollutant removal from wastewaters, including *Opuntia ficus-indica*. This chapter focuses on wastewater treatment strategies involving material parts in sewage containing high levels of chemical oxygen demand and turbidity, heavy metals and pesticides.

Keywords Coagulation · Flocculation · Heavy metal · *Opuntia ficus-indica* · Pesticide · Turbidity · Wastewater

Abbreviations

BOD	Biochemical oxygen demand
COD	Chemical oxygen demand
DDT	Dichloro-diphenyl-trichloroethane
ΔH	Enthalpy change
$FeCl_3$	Ferric chloride
FT-IR	Fourier-transform infrared
ΔG	Gibbs free energy change
OFI	*Opuntia ficus-indica*

© The Author(s) 2018

M. Barbera and G. Gurnari, *Wastewater Treatment and Reuse in the Food Industry*, Chemistry of Foods, https://doi.org/10.1007/978-3-319-68442-0_4

4.1 Removal of Pollutants in Wastewaters and New Strategies. *Opuntia ficus-indica* as Sorbent Material

In recent years, increased industrial and agricultural activities and the correlated population growth led to overexploitation of natural resources [1]; the production of large volumes of sewage has inevitably resulted in the increased generation of various types of pollutants [2]. One of the consequences of this rapid growth is represented by environmental disorders; the demand for clean and safe water has increased tremendously in different sectors [3]. For these reasons, the hazardous pollution of wastewater is one of the most important environmental problems worldwide.

A wide range of treatment technologies are available at present as remediation techniques for wastewaters, in addition to remediation systems mentioned in Chap. 2 [4–6]:

- Chemical precipitation.
- Ion-exchange systems.
- Electrochemical treatments.
- Solvent extraction.
- Coagulation/flocculation methods.[1]
- Adsorption (with activated carbon and zeolites as adsorbents).

Despite the existence of several strategies, these technologies may often show some disadvantages including high sludge volume generation, unsatisfactory removal of pollutants (low concentration), high initial costs, process complexity, high chemical consumption and high maintenance and operation costs [6]. Hence, there is a constant need of efficient, low-cost and alternative wastewater treatments [7].

Recently, the possible use of certain natural substances as removal agents against pollutants in wastewaters in specific sectors—agriculture or food processing—instead of known water and wastewater treatments has been observed. Interestingly, similar options could be very useful when speaking of reuse by-products [8]. A vast amount of biosolid matters have been examined for pollutant removal from wastewaters: one of these possible materials, named *Opuntia ficus-indica* (OFI), has been studied by several authors as biosorbent and coagulant/flocculant agent. Several researchers report the efficient and effective removal of pollutants by OFI in its raw form and as physically or chemically treated material [8, 9]. In particular, OFI cladodes can be used as fresh plant parts or dry powdered materials. This chapter focuses on wastewater treatment strategies involving raw OFI plant parts in sewage containing high chemical oxygen demand (COD) and turbidity values, and notable amounts of heavy metals and pesticides.

---

[1]The following chemicals can be used as coagulation agents: aluminium, ferrous sulphate and ferric chloride. In addition, polyaluminium chloride, polyferric sulphate and polyacrylamide can be considered as flocculants.

4.2 Diffusion and Use of *Opuntia ficus-indica*

The *Opuntia* genus belongs to the *Cactaceae* family with more than 360 species. *OFI* is a tropical or subtropical plant (*Opuntia* spp.) native to the USA, Mexico and South America, but it grows well in other areas, including Africa, Australia and the Mediterranean region [10]. With relation to the Mediterranean basin, some *Opuntia* species (mainly OFI) were introduced five centuries ago from original areas. These plants have a remarkable resistance during prolonged drought periods due to their adaptive features. OFI grows wild in arid or semi-arid countries and is widely cultivated all over the world [11]. This plant represents an important potential source of food and feed in many desert areas. OFI can be used as food thickener [12], food emulsifier [13] and biocoagulant or bioflocculants; other options are known in the pharmaceutical and cosmetic sectors [14]. Moreover, OFI fibres are suitable for low-cost papermaking applications [15] and are also employed in the traditional medicine of several countries [16].

4.3 *Opuntia ficus-indica*—Chemical Features

Opuntia cladodes are well known because of the production of mucilage. This matter is a complex polymeric substance (mixture of carbohydrates) with a reported molecular weight of 2.3×10^4–4.3×10^6 Da [17] and a highly branched structure [18] with variable proportions of L-arabinose, D-galactose, L-rhamnose and D-xylose, as well as galacturonic acid in different proportions according to several authors [19–22]. These molecules are the active components responsible for large adsorption attitude of *Opuntia* cactus. Mucilage also increases osmotic pressure with enhanced water retention capacity [17].

Aside from the above-mentioned main compounds, polygalacturonic acid (Fig. 4.1) is also present. This polymeric substance, derived from galacturonic monomers, has been reported to be the main component responsible for *Opuntia* coagulation activity [8]. In addition, sugars such as L-arabinose and D-galactose (Fig. 4.2), L-rhamnose and D-xylose are relevant compounds, ranging from 74.6 to 92.0% of the total amount of OFI mucilage (Fig. 4.2).

According to the hypothetical structure of mucilage produced by OFI [20], branches of the main chain are formed by three D-galactose units joined to residues

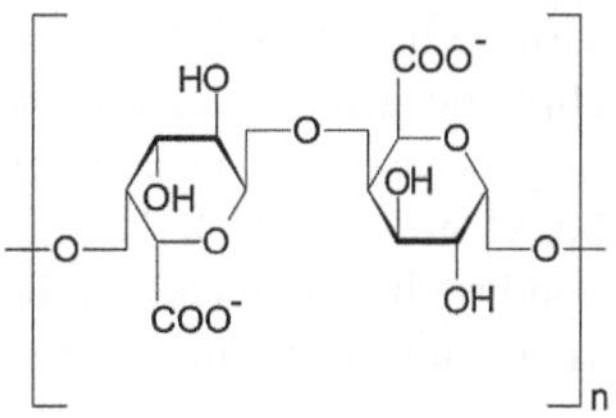

Fig. 4.1 Structure of a polymeric compound found in OFI: poly-α-(1,4)-D-galacturonic acid

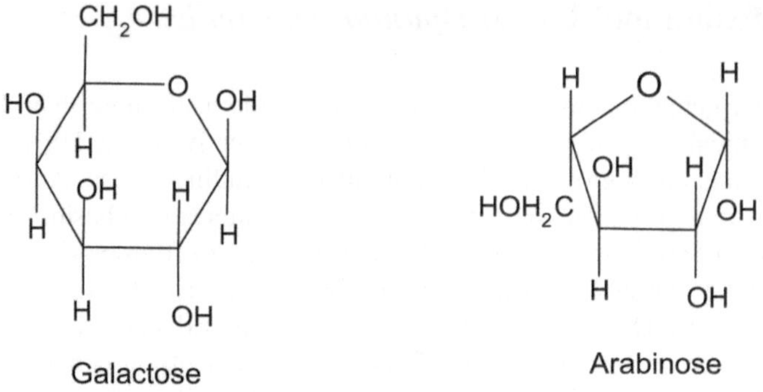

Fig. 4.2 General structure of two sugars commonly found in OFI mucilage: galactose and arabinose. These molecules—L-arabinose and D-galactose—may reach 44.1–45.6% of the total amount of OFI mucilage. In general, the quantity of sugars may range from 74.6 to 92.0% of the total amount of OFI mucilage, including these molecules, L-rhamnose and D-xylose [19–22]

of arabinose and D-xylose. Therefore, L-arabinose is the most abundant sugar in the chemical structure of mucilages. Functional groups of arabinose and D-xylose are well localised spatially when speaking of favoured interactions in an intermolecular form. The observed viscosity could probably be higher under these conditions on condition that mucilage is exposed to water [17].

4.4   FT-IR Characterisation of OFI Cladodes

The Fourier-transform infrared (FT-IR) technique has been used with the aim of exploring the nature of functional groups at the surface of the biosorbent agent.

FT-IR spectroscopy is an important analytical technique able to detect vibration characteristics of chemical functional groups. FT-IR spectra provide a 'chemical fingerprint' of materials by means of the correlation of observed absorption frequencies in the sample with known absorption frequencies for certain bonds (these values are available in dedicated infrared absorption frequency libraries) [23]. Several authors reported peculiar FT-IR spectra of dried prickly pear cactus cladodes in the range of 400–4000 cm^{-1}. With reference to OFI, strong and superimposed bands have been observed in the 3600–3200 cm^{-1} range (overlapping of O–H and N–H stretching vibrations) [17, 24]. Absorption bands at 2921 and 2850 cm^{-1} are indicative of asymmetric stretching vibration for –CH$_2$ groups in carboxylic acids [25] and the symmetric stretching vibration of –CH$_3$ groups in aliphatic acids, respectively [26]. Observed peaks between 1730 and 1710 cm^{-1} may be ascribed to stretching vibration of carbonyl bonds (due to non-ionic carboxyl groups –COOH and –COOCH$_3$) and could be also assigned to hydrogen bonding between carboxylic acids or their esters [27]. Stretching vibration bands at 1650 and 1658 cm^{-1} are due to asymmetric stretching of carboxylic and amidic groups, respectively [11, 28]. Moreover, peaks observed at 1370 cm^{-1} can be assigned to the stretching vibrations of symmetric or asymmetric ionic carboxylic groups (–COOH) of pectin [11]. In addition, absorption peaks around 1155 and 1070 cm^{-1} may be ascribed to P–OH stretching vibrations [11, 27], while the band at 1072 cm^{-1} could reflect the vibration of C–O–C and –OH groups in polysaccharide structures.

The FT-IR analysis indicates that dried OFI surface contains a variety of functional groups such as carboxyl, hydroxyl, sulphate, phosphate, aldehydic, ketonic, carbonyl, amide, amine and alkyl groups. On these bases, it may be inferred that this biomaterial can give good results in terms of efficient reduction, coagulation/flocculation and biosorption of pollutants from wastewaters.

4.5 Application of *Opuntia ficus-indica* in Wastewater Treatments

As above mentioned, there are a variety of ways in which *OFI* can be employed. Several authors [9, 23, 27] reported that OFI cladodes are also used for the treatment of wastewaters (coagulation/flocculation and biosorption processes): cladodes are used either as fresh plant parts or as dry powdered material.

4.5.1 Bioadsorption Treatments

Several researchers reported that adsorption is one of the most effective processes with references to advanced wastewater treatments. Therefore, many industries use adsorption techniques (mainly in the tertiary stage of biological treatment) for reducing hazardous inorganic/organic pollutants present in effluents [6].

The adsorption method refers to a process whereby a material moves from the aqueous or gaseous phase to the solid surface where it is physically and chemically bound [29]. Adsorption by activated carbon represents the most efficient way, but employed materials are highly expensive and regeneration or recycling options are not contemplated. On the other side, biosorption is an emerging technique offering the use of cheap and alternate biological materials to remove substances from solutions. Such matters can be of organic or inorganic nature: they can be found in gaseous, soluble or insoluble forms [30]. Functional groups present in these biomaterials—carboxyl, hydroxyl, sulphydryl and amide groups—make it possible interactions with some pollutant, such as metal ions and pesticides dissolved in waters [7, 27, 28]. The major advantages of biosorption (in comparison with other procedures) are as follows:

1. Lower price.
2. High effectiveness.

3. Availability of materials.
4. Rapidity of the involved process.
5. Reversibility.
6. Regeneration of adsorbent agents by means of suitable desorption process (chemical or biological sludge is minimised).

For these reasons, biosorption process is one of the most widely used methods for the removal of pollutants from wastewater [6, 31]. As a consequence, the research of alternative adsorbent materials in wastewater treatment is gaining prominence. Recently, the survey of new biomaterials has received the greatest attention for the removal of both inorganic and organic pollutants. Numerous works have been published with a primary goal: the investigation of removal of different pollutants (either in gas or liquid medium) using adsorbent materials such as agricultural and industrial wastes (peanut hull, peanut husk, eggshells, lignite, by-products of the production chain for olive oils) [32–34], fungi [35], bacteria [36], crustacean shells, clay and peat moss [34].

Generally, basic criteria for these potential adsorbents (with relation to wastewater treatments) are based on adsorption equilibria and kinetics [37]. Mechanistic modelling of kinetic parameters plays a crucial role with concern to the evaluation of adsorption performances for a given compound and target contaminants. On the other side, thermodynamic aspects are important in terms of assessing the feasibility of adsorption reactions as well as the stability of solid–liquid-phase systems.

The nature of adsorption process can be described by means of thermodynamic parameters including enthalpy change (ΔH) and Gibbs free energy change (ΔG) [33]. Hence, the mechanistic modelling of kinetics and thermodynamic parameters would provide a substantial understanding to ensure the removal efficiency of adsorbent materials in wastewaters.

4.5.2   Kinetics Adsorption

Kinetics adsorption describes the solute uptake rate, which in turn governs residence time and reaction pathways of the adsorption process. Kinetic data are derived from the variation of pollutants removed per given time (q_t) against time (t) [38, 39]. Kinetic modelling not only allows the estimation of adsorption rates but also leads to suitable rate expressions characteristic of possible reaction mechanisms.

The most prevalent kinetic models investigated from several authors are the pseudo-first-order and pseudo-second-order kinetic models [37, 40]. The pseudo-first-order rate expression, based on solid capacity, is generally expressed by Eq. 4.1

$$\frac{dq_t}{dt} = k_1(q_e - q_t) \tag{4.1}$$

where q_e is the amount of adsorbed material at equilibrium (mg g^{-1}), q_t is the amount adsorbed at the time t (mg g^{-1}), k_1 is the constant rate constant of first-order adsorption (min^{-1}) and t is the contact time (min) [11]. The linearised expression is expressed by Eq. 4.2.

$$\log(q_e - q_t) = \log q_e - k_1 t \tag{4.2}$$

The constant k_1 can be determined from the slope of the plots of log $(q_e - q_t)$ versus t.

The pseudo-second-order model is based on the assumption that the adsorption follows a second-order chemisorption, as shown by Eq. 4.3

$$\frac{dq_t}{dt} = k_2(q_e - q_t)^2 \tag{4.3}$$

where k_2 is the rate constant of second-order adsorption (g mg^{-1} min). In this model, the rate-limiting step is the surface adsorption that involves chemisorption, where the removal from a solution is due to physicochemical interactions between the two phases. In reactions involving chemisorption of adsorbate on a solid surface without desorption of products, adsorption rate decreases with time due to an increased surface coverage [11]. The linearised form of the pseudo-second-order kinetic model can be expressed as follows (Eq. 4.4):

$$\frac{t}{q_t} = \frac{1}{k_2(q_e)^2} + \frac{t}{q_e} \tag{4.4}$$

where k_2 values were determined from the slope of the plots of t/q_t against t. The remarkable advantage of this model is correlated with the accuracy in the description of the whole kinetic experimental data [41].

4.5.3 Adsorption Equilibria

The adsorption model is a useful tool giving information about the theoretical maximum adsorption capacity and possible interactions between adsorbents and adsorbate [7]. Adsorption isotherms are equilibrium relationships between the quantity of adsorbate per unit of adsorbent (q_{eq}) and its equilibrium solution concentration (C_{eq}) [38, 39]. Several available equations or models describe this function: the most part of works published in relation to adsorption adopt either the Langmuir or Freundlich isotherm (or both) for adsorption data correlation [33, 42].

The Langmuir adsorption isotherm represents the equilibrium distribution of metal ions between the solid and liquid phases. This relation is valid for dynamic equilibrium adsorption–desorption processes on completely homogeneous surfaces with negligible interactions between adsorbed molecules. In other terms, the Langmuir adsorption isotherm describes quantitatively the formation of a mono-layer adsorbate on the outer surface of adsorbents, with the assumption that all binding sites have equal affinity for sorbate and the sorption takes place at specific homogeneous sites within the adsorbent [43]. Although this description gives no information about the mechanism, it is still used to obtain the uptake capacities of sorbents. Langmuir isotherm is shown as follows (Eq. 4.5):

$$q_{eq} = \frac{Q_{max} k_L C_{eq}}{\left(1 + k_L C_{eq}\right)} \tag{4.5}$$

Langmuir adsorption parameters were determined by transforming the Langmuir equation into linear form (Eq. 4.6).

$$\frac{C_{eq}}{q_{eq}} = \frac{C_{eq}}{q_{max}} + \frac{1}{k_L q_{max}} \tag{4.6}$$

where C_{eq} is the equilibrium concentration of adsorbate (mg L^{-1}), q_{eq} is the amount of metal adsorbed per gram of adsorbent at the equilibrium (mg g^{-1}), q_{max} is the maximum monolayer coverage capacity (mg g^{-1}) and k_L is the Langmuir isotherm constant. Basic terms of the linearised equation may be computed from the slope and intercept of the Langmuir plot of C_{eq}/q_{eq} versus C_{eq} [7, 11].

The Freundlich isotherm is commonly used to describe adsorption features for the heterogeneous surface: it assumes that adsorption energy varies as a function of surface coverage. This equation is also applicable to multilayer adsorption and is expressed by the following Eq. 4.7 [11]:

$$Q_{eq} = k_F C_{eq}^{1/n} \tag{4.7}$$

where k_F is the Freundlich isotherm constant, n is the adsorption intensity, C_{eq} is the equilibrium concentration of adsorbate (mg L^{-1}) and Q_{eq} is the amount of metal adsorbed per gram of the adsorbent at equilibrium (mg g^{-1}). This relation can be shown in a linearised equation as follows (Eq. 4.8):

$$\log Q_{eq} = \frac{1}{n} \log C_{eq} + \log k_F \tag{4.8}$$

k_F and n are parameters characteristic of the sorbent–sorbate system: they must be determined by data fitting. Consequently, linear regression is generally used to determine parameters of kinetic and isotherm models.

In particular, the constant k_F is an approximate indicator of adsorption capacity, while $1/n$ is a function of the strength of adsorption in the adsorption process [44].

If $n = 1$, the partition between the two phases are independent of the concentration on condition that $1/n$ is <1 (normal adsorption). On the other hand, should the term $1/n$ be >1, a cooperative adsorption would be assumed. This expression can be reduced to a linear adsorption isotherm when $1/n = 1$. Should n lies in the range 1–10, the expression would indicate a favourable sorption process [45].

4.5.4   Factors that Influence the Adsorption Phenomenon

Adsorption mechanisms involve the outer surface and can be variegated due to chemical–physical features of the specific surface area (particle size and functional groups, heterogeneous reactive sites). A good adsorbent material should generally possess a porous structure (resulting in high surface area) and the time taken for adsorption equilibrium to be established should be as small as possible, so that it can be used to remove pollutants in a reduced time [3]. Furthermore, physicochemical conditions under which the biosorption takes place and the environmental conditions such as pH and temperature strongly influence the adsorption process [46].

4.5.4.1   Surface Area, pH and Temperature

The capacity of the adsorbent material is strongly related to the extension of the specific surface area, the structure and chemical nature. These parameters control swelling properties and the diffusion in the polysaccharide matrix and affect its features [48, 49]. The greater the surface area of a specific biosorbent, the greater the substance biosorption, provided that all other parameters influencing the process are kept constant.

In general, the efficiency of adsorption is strongly dependent on the particle size of the adsorbent agent. This is due to the fact that the smaller particle determines a larger surface area of adsorbent materials on a macroscopic scale, thus increasing the number of adsorption sites and enhancing adsorption capacity [7, 50]. pH of the aqueous solution is one of the major parameters controlling the biosorption process. In fact, it strongly influences the biosorption availability of present ions; it determines the availability of Lewis basic sites, and it also defines the speciation of metal ions. Moreover, pH controls the protonation of different surface functional groups [8].

Temperature is found to be an important parameter influencing the thermodynamics of the biosorption process. In fact, the change in temperature causes a change in thermodynamic parameters like ΔG, ΔH and entropy change. These parameters contribute to the comprehension of the sorption mechanism; also, they are directly related to the variation of kinetic energy, thus influencing the diffusion process [51, 52].

4.6   Application of *Opuntia ficus-indica* as Biosorbent Material in Wastewater Treatments

As above mentioned, the use of natural biomaterials is a promising alternative due to their relative abundance and their low commercial value. Several authors tested fresh or dry OFI as biosorbent material to remove metal ions and pesticides from wastewaters. With reference to dry materials, spines are removed and OFI cladodes are washed with water, cut and dried. On the other hand (fresh material), cladodes are peeled, macerated on the entire pads and then refrigerated [8].

4.6.1   Removal of Pesticides

The use of pesticides in agricultural practices worldwide has increased dramatically during the last two decades [53]. In particular, they represent a strong problem in developing countries due to weak regulation and the high cost of water treatment systems [54]. As a result of the widespread use of pesticides, decontamination of water resources by pesticide residues is one of the major challenges for the preservation and sustainability of the environment [55, 56]. The potentiality of OFI as biosorbent material to remove pollutants from surface waters has been evaluated [54]. In particular, researchers tested the efficiency of fresh and dry OFI in batch and column systems to eliminate pesticides aldrin, dieldrin and dichloro-diphenyl-trichloroethane (DDT). In particular, these researchers found that the remarkable pesticides adsorption on dry and fresh OFI is apparently dependent on the particle size of adsorbent materials and the highest removal percentage. In particular [54]:

(a) Fresh OFI materials can remove aldrin, dieldrin and DDT with acceptable results—19.1 to 42.6, 28.7 to 69.4 and 5.2 to 10.5%, respectively—depending on particle sizes (ranging from 3 to 1 cm). Best results are obtained with the smaller particle dimension.
(b) On the other hand, dry OF can show ameliorated performances depending on particle sizes (ranging from <0.25 to 1.0–2.0 mm). Substantially, the smallest particle sizes allow excellent pesticide removals. For example, DDT is removed up to 99.2% for particles of diameter <0.25 mm, while 1.0–2.0 mm—particles can remove up to 77.1% for this pesticide.

4.6.2   Removal of Metal Ions

Heavy metals are present in virtually every aspect of modern consumerism such as construction materials, cosmetics, medicines, processed foods, fuel sources, appliances and various personal care products (3–5). The human exposure to any of the many harmful heavy metals prevalent in our environment is apparently unavoidable. Many heavy metals are known to be significantly toxic: these elements are not biodegradable and tend to accumulate themselves in living organisms with damages in humans and living species even in low concentration. For these reasons, their removal from effluents or the reduction in concentration are needed at present; the removal and recovery of heavy metals from wastewater are significant when speaking of protection of the environment and human health [7].

In aqueous solutions, metal ions are present under various chemical forms depending on environmental conditions such as temperature, pH, ionic strength and the chemical composition of the medium. With reference to the adsorption mechanism of metal ions from aqueous solutions, the speciation of metal ions and hence the pH of the aqueous medium are reported to play a dominant role. A perusal of available literature reveals that the optimum pH at which maximum adsorption capacity is achieved depends ultimately on the nature of metal ion, irrespective of physical characteristics of adsorbent materials or its precursor source [57, 58]. In general, it is possible to affirm that the amount of metal ions adsorbed is low at lower pH values because large quantities of hydrogen ions compete with metal cations for biomass surface. As the pH increased, the number of negatively charged sites increases, with consequent enhanced biosorption of positively charged metal cations through electrostatic attraction forces [59, 60]. Some authors studied the removal/adsorption capacity of metal ions with OFI. Table 4.1 shows some available bioadsorbents, including *Opuntia* materials, used with their respective better adsorption capacity in removing metal ions from water. Results of these investigations by different papers and approximate comparisons with other adsorbing agents (Table 4.1) show that *Opuntia* materials may be good biosorbent material for heavy metal removal (Fig. 4.3).

4.6.3 *Other* Opuntia ficus-indica *Applications*

OFI materials are also investigated to evaluate the potentiality to pollutant removal from tannery wastewater. Swathi and co-workers [61] reported that cactus powder can be used effectively as an adsorbent for pre-treatment for tannery wastewater. They found that pollutant parameters such as turbidity, biochemical oxygen demand (BOD), COD, chromium, iron, sulphate and chloride were reduced to the following levels, respectively, 70.9, 57.2, 64.3, 67.4, 98, 86.2 and 83.2%.

Table 4.1 Adsorption capacities (mg/g) of OFI materials and some bioadsorbents for the removal of selected metal ions [7, 11, 31, 32, 34–36, 61–64, 66–71]

Adsorbing agent	Removed metals (mg/g)						Reference
	Cadmium	Copper	Lead	Chromium	Hexavalent chromium	Zinc	
Raw cladodes				2251.5			[61]
Raw fibres		41.3					[62]
Raw ectodermis				11.7		5.7	[63]
Raw cladodes	30.4		98.6		.		[11]
Raw cladodes					18.5		[31]
Raw ectodermis					16.4		[31]
Rice husk carbon					45.6		[64]
Red mud	10.6	19.7				13	[64]
Phomopsis sp.	26	25	179				[35]
Bacillus lentus	30	30					[65]
Rizopus arrhizus			104				[36]
Triticum aestivum	51.58	17.4	87	93	40.8	16.4	[7]
Streptomyces rimosus			135				[66]
Clorella mintissima		11	10				[67]
Blast furnace slag			40		7.5		[64]
Blast furnace sludge	10.1	23.7	79.9	10		9.6	[64]
Olive cake	6.5		30		33.4		[32]
Lignin			1865				[68]
Waste slurry			1030				[69]
Clinoptilolite	70	2	62				[70, 71]
Chabazite	137		175				[70]
Chitosan	6	222	16			75	[34, 64]

Fig. 4.3 An approximate comparison between three different adsorbing materials against heavy metals in waters. The removal/adsorption performance of metal ions with OFI materials has been studied (Table 4.1). The comparison between average data for OFI materials (the highest result has been used for comparison) and other available bioadsorbents show that *Opuntia* materials may be good biosorbent material for heavy metal removal

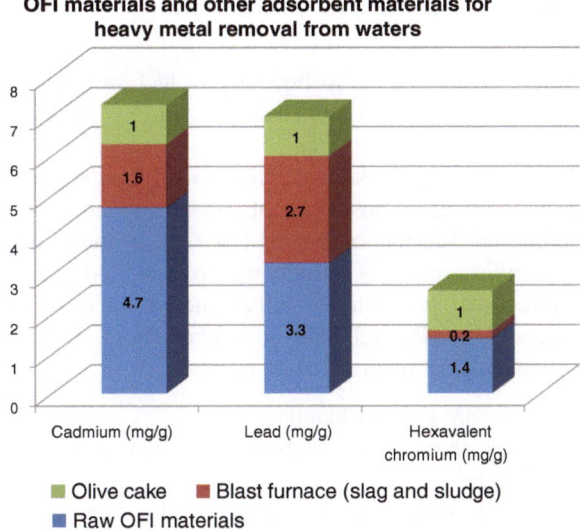

4.7   Application of *Opuntia ficus-indica* as Coagulant/Flocculant Material in Wastewater Treatments

Several studies have reported the examination of coagulation and flocculation processes for the treatment of different kinds of industrial wastewaters like tannery, textile and food processing [8, 72]. Coagulation and flocculation are commonly used treatments to remove colloidal particles from water and wastewater. In these processes, coagulant/flocculant compounds are added to wastewater in order to destabilise colloidal materials. The addition of coagulant is due to the aggregation of colloidal particles through neutralisation of forces keeping them apart, while added flocculants promote aggregation of particles into large agglomerates which can be physically separated from the liquid phase by floatation, settling or adsorption [6, 73]. At present, coagulant/flocculant compounds such as ferric chloride and/or synthetic polymers are the most applied substances because of their efficiency, but their use is accompanied by large consumption of chemicals that leads to production of large volume of non-biodegradable sludge. Therefore, recent developments have been made possible by means of the use of natural organic polymers and polyelectrolytes as flocculants and/or coagulants in wastewater treatments.

4.7.1   Flocculation Treatments

Recent developments in flocculation technology have proposed the use of natural organic polymers as flocculants and/or flocculation aids in river water and wastewater treatments, taking precedence over inorganic and synthetic polymers. Recently, researchers have concentrated their studies on the flocculation/adsorption technology using low-cost, abundant and non-conventional materials instead of traditional flocculation agents [74, 75]. Natural coagulants/flocculants are considered environmentally friendly due to their biodegradability, stability and low cost; it is forecasted that their use will promote more biodegradable sludge at the end of the process [18]. Renault and co-workers have highlighted the advantages of these polymeric flocculants [76] as follows:

(a) They are easy to handle.
(b) They show high solubility in water and a promising reduction of sludge volume.
(c) These compounds are readily available and biodegradable, and they produce large, dense and compact flocs with good settling characteristics.

At present, a number of natural polymers and polyelectrolytes have been explored and evaluated to be effective in wastewater decontamination either through adsorption or coagulation/flocculation processes. Many authors have also studied the efficiency of OFI cladodes as eco-friendly flocculants/coagulants [77, 78]. As mentioned before, OFI was defined a cheap and easily available plant and FT-IR studies confirmed the presence of various functional groups responsible for the coagulation/flocculation process. In fact, OFI cladodes are mainly constituted by heteropolysaccharides [79, 80].

Some of the reported results by different authors about OFI performances and also the results of comparatives studies with commercial flocculants are briefly shown in this section.

Bouatay and co-workers [18] have investigated the OFI performance in flocculation processes compared with two commercial flocculants (a cationic flocculant, EPENWATE EXP 31/1 and the anionic agent polyacrylamide $A_{100}PWG$). These authors evaluated the decolourisation, the COD removal and the turbidity abatement. They demonstrated that OFI mucilage had a better performance and, in particular, performance of the cactus mucilage was higher than the achieved by EPENWATE EXP 31/1 and equal to obtained by the $A_{100}PWG$. On these bases, authors inferred that the obtained flocs using cactus mucilage and the $A_{100}PWG$ as flocculants are bigger and heavier than those arising from the system based on EPENWATE EXP 31/1.

Torres and co-workers [81] have investigated the OFI efficiency in comparison with two biopolymers (*Prosopis galactomannan* and *Opuntia* mucilage) and a chemical coagulant very frequently used in wastewater treatment: ferric chloride ($FeCl_3$), with relation to COD, salt and turbidity diminution from municipal wastewaters. These authors found that COD, salt and turbidity diminution using biopolymers were comparable to those found when using $FeCl_3$. They also reported that the sludge production was in general lower for both the biopolymers instead of $FeCl_3$, though it was very dependent on pH and the amount of employed coagulant/flocculant agents.

Miller and co-workers [82] and Pichler and co-workers [83] explored the use of *Opuntia* materials in turbidity removal, respectively, in synthetic clay solution and drinking water. Both groups found that OFI performances are good when speaking of turbidity diminution in wastewater. Also, Pichler and co-workers [83] demonstrated that the OFI gelling extract was a good competitor as a flocculant agent. In all cases, the examined OFI appeared to be a viable, cheap, eco-friendly and effective natural alternative when applied as coagulant/flocculant and adsorbent material in wastewater treatments. Hence, the exploitation of this abundant, renewable, simple and non-toxic natural resource would be encouraged.

References

1. Gupta VK, Nayak A (2012) Cadmium removal and recovery from aqueous solutions by novel adsorbents prepared from orange peel and Fe_2O_3 nanoparticles. Chem Eng J 180:81–90. doi:10.1016/j.cej.2011.11.006
2. Bhatnagar A, Minocha AK (2006) Conventional and non-conventional adsorbents for removal of pollutants from water-a review. Indian J Chem Technol 13:203–217
3. Gupta VK (2009) Application of low-cost adsorbents for dye removal—a review. J Environ Manage 90(8):2313–2342. doi: 10.1016/j.jenvman.2008.11.017
4. Gupta VK, Kumar R, Nayak A, Saleh TA, Barakat MA (2013) Adsorptive removal of dyes from aqueous solution onto carbon nanotubes: a review. Adv Colloid Interface Sci 193:24–34. doi:10.1016/j.cis.2013.03.003
5. Srinivasan A, Viraraghavan T (2010) Decolorization of dye wastewaters by biosorbents: a review. J Environ Manage 91(10):1915–1929. doi:10.1016/j.jenvman.2010.05.003
6. Fu F, Wang Q (2011) Removal of heavy metal ions from wastewaters: a review. J Environ Manage 92(3):407–418. doi:10.1016/j.jenvman.2010.11.011
7. Farooq U, Kozinski JA, Khan MA, Athar M (2010) Biosorption of heavy metal ions using wheat based biosorbents-a review of the recent literature. Bioresour Technol 101(14):5043–5053. doi:10.1016/j.biortech.2010.02.030
8. Nharingo T, Moyo M (2016) Application of *Opuntia ficus-indica* in bioremediation of wastewaters. A critical review. J Environ Manage 166:55–72. doi:10.1016/j.jenvman.2015.10.005
9. Vishali S, Karthikeyan R (2015) Cactus *Opuntia (ficus-indica)*: an eco-friendly alternative coagulant in the treatment of paint effluent. Desalin Water Treat 56(6):1489–1497. doi:10.1080/19443994.2014.945487
10. De Leo M, De Abreu MB, Pawlowska AM, Cioni PL, Braca A (2010) Profiling the chemical content of *Opuntia ficus-indica* flowers by HPLC–PDA-ESI-MS and GC/EIMS analyses. Phytochem Lett 3(1):48–52. doi:10.1016/j.phytol.2009.11.004
11. Barka N, Ouzaouit K, Abdennouri M, El Makhfouk M (2013) Dried prickly pear Cactus (*Opuntia ficus-indica*) cladodes as a low-cost and eco-friendly biosorbent for dyes removal from aqueous solutions. J Taiwan Inst Chem Eng 44(1):52–60. doi:10.1016/j.jtice.2012.09.007
12. Cardenas A, Higuera-Ciapara I, Goycoolea FM (1997) Rheology and aggregation of cactus (*Opuntia ficus indica*) mucilage in solution. J Prof Assoc Cactus Dev 2:152–157

13. Medina-Torres L, Brito-De La Fuente E, Torrestiana-Sanchez B, Katthain R (2000) Rheological properties of the mucilage gum (*Opuntia ficus indica*). Food Hydrocoll 14(5): 417–424. doi:10.1016/S0268-005X(00)00015-1

14. Abdel-Hameedn ESS, Nagaty MA, Salman MS, Bazaid SA (2014) Phytochemicals, nutritionals and antioxidant properties of two prickly pear cactus cultivars (*Opuntia ficus indica* Mill.) growing in Taif, KSA. Food Chem 160:31–33. doi:10.1016/j.foodchem.2014. 03.060

15. Mannai F, Ammar M, Yanez JG, Elaloui E, Moussaoui Y (2016) Cellulose fiber from Tunisian Barbary Fig "*Opuntia ficus-indica*" for papermaking. Cellulose 23(3):2061–2072. doi:10.1007/s10570-016-0899-9

16. Agozzino P, Avellone G, Ceraulo L, Ferrugia M, Filizzola F (2005) Volatile profile of sicilian prickly pear (*Opuntia ficus-indica*) by SPME-GC/MS analysis. Ital J Food Sci 17(3):341–348

17. Rodríguez-González S, Martínez-Flores HE, Chávez-Moreno CK, Macías-Rodríguez L, Zavala-Mendoza E, Garnica-Romo MG, Chacón-García L (2014) Extraction and characterization of mucilage from wild species of *Opuntia*. J Food Process Eng 37(3):285–292. doi:10. 1111/jfpe.12084

18. Bouatay F, Mhenni F (2014) Use of the cactus cladodes mucilage (*Opuntia Ficus Indica*) as an eco-friendly flocculants: process development and optimization using stastical analysis. Int J Environ Res 8(4):1295–1308

19. Amin ES, Awad OM, El-Sayed MM (1970) The mucilage of *Opuntia ficus indica* mill. Carbohydr Res 15(1):159–161. doi:10.1016/S0008-6215(00)80304-3

20. Nobel P, Cavelier J, Andrade JL (1992) Mucilage in cacti: its apoplastic capacitance, associated solutes, and influence on tissue 5. J Exp Bot 43(250):641–648. doi:10.1093/jxb/43. 5.641

21. Fox DI (2011) Cactus mucilage-assisted heavy metal separation: design and implementation. Dissertation, University of South Florida

22. Trachtenberg S, Mayer AM (1981) Composition and properties of *Opuntia ficus-indica* mucilage. Phytochem 20(12):2665–2668. doi:10.1016/0031-9422(81)85263-6

23. Fox DI, Pichler T, Yeh DH, Alcantar NA (2012) Removing heavy metals in water: the interaction of Cactus mucilage and arsenate (As (V)). Environ Sci Technol 46(8):4553–4559. doi:10.1021/es2021999

24. Sarı A, Tuzen M (2009) Kinetic and equilibrium studies of biosorption of Pb(II) and Cd(II) from aqueous solution by macrofungus (*Amanita rubescens*) biomass. J Hazard Mater 164 (2):1004–1011. doi:10.1016/j.jhazmat.2008.09.002

25. Díaz KDM, Reyes TF, Cabrera P, Sánchez MD, García MA, de Posada Piñán E (2013) Characterization of laser-treated Opuntia using FT-IR spectroscopy and thermal analysis. Appl Phys A Mater Sci Process 112(1):221–224. doi:10.1007/s00339-012-7308-5

26. Ishurd O, Zgheel F, Elghazoun M, Elmabruk M, Kermagi A, Kennedy JF, Knill CJ (2010) A novel (1-4)-a-D-glucan isolated from the fruits of *Opuntia ficus-indica* (L.) Miller. Carbohydr Polym 82(3):848–853. doi:10.1016/j.carbpol.2010.06.006

27. Betatache H, Aouabed A, Drouiche N, Lounici H (2014) Conditioning of sewage sludge by prickly pear Cactus (*Opuntia ficus-indica*) juice. Ecol Eng 70:465–469. doi:10.1016/j. ecoleng.2014.06.031

28. Jadhav MV, Mahajan YS (2014) Assessment of feasibility of natural coagulants in turbidity removal and modelling of coagulation process. Desalin Water Treat 52(31–33):5812–5821. doi:10.1080/19443994.2013.816875

29. Barakat MA (2011) New trends in removing heavy metals from industrial wastewater. Arab J Chem 4(4):361–377. doi:10.1016/j.arabjc.2010.07.019

30. Ghasemi M, Naushad M, Ghasemi N, Khosravi-fard Y (2014) A novel agricultural waste based adsorbent for the removal of Pb(II) from aqueous solution: kinetics, equilibrium and thermodynamic studies. J Ind Eng Chem 20(2):454–461. doi:10.1016/j.jiec.2013.05.002

31. Fernández-López JA, Angosto JM, Avilés MD (2014) Biosorption of hexavalent chromium from aqueous medium with opuntia biomass. Sci World J 670249. doi:10.1155/2014/67024

32. Anastopoulos I, Massas I, Constantinos E (2015) Use of residues and by-products of the olive-oil production chain for the removal of pollutants from environmental media: a review of batch biosorption approaches. J Environ Sci Heal Part A 50(7):677–718

33. Anastopoulos I, Kyzas GZ (2014) Agricultural peels for dye adsorption: a review of recent literature. J Mol Liq 200:381–389. doi:10.1016/j.molliq.2014.11.006

34. Babel S, Kurniawan TA (2003) Low-cost adsorbents for heavy metals uptake from contaminated water: a review. J Hazard Mater 97(1–3):219–243. doi:10.1016/S0304-3894 (02)00263-7

35. Saiano F, Ciofalo M, Cacciola SO, Ramirez S (2005) Metal ion adsorption by *Phomopsis* sp. biomaterial in laboratory experiments and real wastewater treatments. Water Res 39 (11):2273–2280. doi:10.1016/j.watres.2005.04.022

36. Selatnia A, Boukazoula A, Kechid N, Bakhti MZ, Chergui A, Kerchich Y (2004) Biosorption of lead (II) from aqueous solution by a bacterial dead *Streptomyces rimosus* biomass. Biochem Eng J 19(2):127–135

37. Doke KM, Khan EM (2013) Adsorption thermodynamics to clean up wastewater; critical review. Rev Environ Sci Biotechnol 12:25–44. doi:10.1016/j.bej.2003.12.007

38. Nharingo T, Shoniwa V, Hunga O, Shumba M (2013) Exploring the biosorption of Methylene Blue dye onto acid treated sugarcane bagasse. Int J Curr Res 5:2169–2175

39. Mahamadi C, Nharingo T (2007) Modelling the kinetics and equilibrium properties of cadmium biosorption by river green alga and water hyacinth weed. Toxicol Environ Chem 89 (2):297–305. doi:10.1080/02772240601010063

40. Kyzas GZ, Matis KA (2014) New biosorbent materials: selectivity and bioengineering insights. Processes 2(2):419–440. doi:10.3390/pr2020419

41. Al-Garni SM (2005) Biosorption of lead by gram-ve capsulated and noncapsulated bacteria. Water SA 31(3):345–350. doi:10.4314/wsa.v31i3.5224

42. Bharathi KS, Ramesh ST (2013) Removal of dyes using agricultural waste as low-cost adsorbents: a review. Appl Water Sci 3(4):773–790. doi:10.1007/s13201-013-0117-y

43. Langmuir I (1918) The adsorption of gases on plane surfaces of glass, mica and platinum. J Am Chem Soc 40:1361–1403

44. Voudrias E, Fytianos F, Bozani E (2002) Sorption–desorption isotherms of dyes from aqueous solutions and wastewaters with different sorbent materials. Global Nest Int J 4(1): 75–83

45. Freundlich H (1906) Over the adsorption in solution. J Phys Chem 57:385–470

46. Volesky B, Holant ZR (1995) Biosorption of heavy metals. Biotechnol Prog 11(3):235–250. doi:10.1021/bp00033a001

47. Aksu Z, Sag Y, Kutsal T (1992) The biosorpnon of copperod by *C. vulgaris* and *Z. ramigera*. Environ Technol 13(6):579–586. doi:10.1080/09593339209385186

48. Tsezos M, Remoundaki E, Hatzikioseyian A (2006) Biosorption-principles and applications for metal immobilization from waste-water streams. In: Proceedings of EU-Asia workshop on clean production and nanotechnologies, Seoul, Korea, 25–26 Oct 2006, pp 23–33

49. Ramachandra TV, Ahalya N, Kanamadi RD (2003) Biosorption of heavy metals. Res J Chem Environ 7(4):71–79

50. Gupta VK, Jain CK, Ali I, Chandra S, Agarwal S (2002) Removal of lindane and malathion from wastewater using bagasse fly ash—a sugar industry waste. Water Res 36(10):2483–2490. doi:10.1016/S0043-1354(01)00474-2

51. Horsfall MJ, Spiff AI (2005) Effects of temperature on the sorption of Pb^{2+} and Cd^{2+} from aqueous solution by caladium bicolor (wild cocoyam) biomass. Electron J Biotechnol 8(2): 162–169. doi:10.4067/S0717-34582005000200005

52. Sawalha MF, Peralta-Videa JR, Romero-Gonzalez J, Gardea-Torresdey JL (2006) Biosorption of Cd(II), Cr(III), and Cr(VI) by saltbush (*Atriplex canescens*) biomass: thermodynamic and isotherm studies. J Colloid Interface Sci 300(1):100–104. doi:10.1016/j.jcis.2006.03.029

53. Diez MC (2010) Biological aspects involved in the degradation of organic pollutants. J Soil Sci Plant Nutr 10(3):244–267. doi:10.4067/S0718-95162010000100004

54. Gebrekidan A, Nicolai H, Vincken L, Teferi M, Asmelash T, Dejenie T, Zerabruk S, Gebrehiwet K, Bauer H, Deckers J, Luis P, De Meester L, van der Bruggen B (2013) Pesticides removal by filtration over cactus pear leaves: a cheap and natural method for small-scale water purification in semi-arid regions. CLEAN Soil Air Water 41(3):235–243. doi:10.1002/clen.201200042

55. Abballe A, Ballard TJ, Dellatte E, di Domenico A, Ferri F, Fulgenzi ARa, Grisanti G, Iacovella N, Ingelido AM, Malisch R, Miniero R (2008) Persistent environmental contaminants in human milk: concentrations and time trends in Italy. Chemosphere 73(1): S220–S227. doi:10.1016/j.chemosphere.2007.12.036

56. Weichenthal S, Moase C, Chan P (2010) Review of pesticide exposure and cancer incidence in the Agricultural Health Study cohort. Environ Health Perspect 118(8):1117–1125

57. Farooq U, Khan MA, Athar M, Kozinski JA (2011) Effect of modification of environmentally friendly biosorbent wheat (*Triticum aestivum*) on the biosorptive removal of cadmium(II) ions from aqueous solution. Chem Eng J 171(2):400–410. doi:10.1016/j.cej.2011.03.094

58. Akhtar M, Iqbal S, Kausar A, Bhanger MI, Shaheen MA (2010) An economically viable method for the removal of selected divalent metal ions from aqueous solutions using activated rice husk. Colloids Surf B 75(1):149–155. doi:10.1016/j.colsurfb.2009.08.025

59. Barka N, Abdennouri M, Makhfouk MEL (2011) Removal of methylene blue and eriochrome black T from aqueous solutions by biosorption on *Scolymus hispanicus L.*: kinetics, equilibrium and thermodynamics. J Taiwan Inst Chem Eng 42:320–326. doi:10.1016/j.jtice.2010.07.004

60. Wang XJ, Xia SQ, Chen L, Zhao JF, Chovelon JM, Nicole JR (2006) Biosorption of cadmium(II) and lead(II) ions from aqueous solutions onto dried activated sludge. J Environ Sci 18(5):840–844. doi:10.1016/S1001-0742(06)60002-8

61. Swathi M, Sathya SA, Aravind S, Sudhakar PK, Gobinath R, Devi DS (2014) Experimental studies on tannery wastewater using Cactus powder as an adsorbent. Int J Appl Sci Eng Res 3(2):436–446. doi:10.6088/ijaser.030200014

62. Prodromou M, Pashalidis I (2013) Copper(II) removal from aqueous solutions by adsorption on non-treated and chemically modified Cactus fibres. Water Sci Technol 68(11):2497–2504. doi:10.2166/wst.2013.535

63. Barrera H, Ureña-Núñez F, Bilyeu B, Barrera-Díaz C (2006) Removal of chromium and toxic ions present in mine drainage by *Ectodermis* of *Opuntia*. J Hazard Mater 136:846–853. doi:10.1016/j.jhazmat.2006.01.021

64. Bhatnagar A, Minocha AK (2006) Conventional and non conventional adsorbents for removal of pollutants from water—a review. Indian J Chem Technol 13(3):203–217. Available http://nopr.niscair.res.in/handle/123456789/7020. Accessed 03 Apr 2017

65. Vianna LNL, Andrade MC, Vicoli JR (2000) Screening of waste biomass from *Saccharomyces cerevisiae*, *Aspergillus oryzae* and *Bacillus lentus* fermentations for removal of Cu, Zn and Cd by biosorption. World J Microb Biotechnol 16(5):437–440. doi:10.1023/A:1008953922144

66. Selatnia A, Madani A, Bakhti MZ et al (2004) Biosorption of Ni^{2+} from aqueous solution by a NaOH-treated bacterial dead *Streptomyces rimosus* biomass. Miner Eng 17(7):903–911. doi:10.1016/j.mineng.2004.04.002

67. Roy D, Greenlaw PN, Shane BS (1993) Adsorption of heavy metals by green algae and ground rice hulls. J Environ Sci Health A 28(21):37–50. doi:10.1080/10934529309375861

68. Srivastava SK, Singh AK, Sharma A (1994) Studies on the uptake of lead and zinc by lignin obtained from black liquor—a paper industry waste material. Environ Technol 15(4):353–361. doi:10.1080/09593339409385438

69. Srivastava SK, Tyagi R, Pant N (1989) Adsorption of heavy metal ions on carbonaceous material developed from the waste slurry generated in local fertilizer plants. Water Res 23 (9):1161–1165. doi:10.1016/0043-1354(89)90160-7

70. Ouki SK, Cheeseman CR, Perry R (1993) Effects of conditioning and treatment of chabazite and clinoptilolite prior to lead and cadmium removal. Environ Sci Technol 27(6):1108–1116

71. Zamzow MJ, Eichbaum BR, Sandgren KR, Shanks DE (1990) Removal of heavy metals and other cations from waste water using Zeolites. Sep Sci Technol 25(13–15):1555–1569. doi:10.1080/01496399008050409

72. De Sena RF, Moreira RF, José HJ (2008) Comparison of coagulants and coagulation aids for treatment of meat processing wastewater by column flotation. Bioresour Technol 99(17): 8221–8225. doi:10.1016/j.biortech.2008.03.014

73. Amuda OS, Amoo IA, Ajayi OO (2006) Performance optimization of coagulant/ flocculant in the treatment of wastewater from a beverage industry. J Hazard Mater 129(1–3):69–72. doi:10.1016/j.jhazmat.2005.07.078

74. Theodoro JDP, Lenz GF, Zara RF, Bergamasco R (2013) Coagulants and natural polymers: perspectives for the treatment of water. Plast Polym Technol 2(3):55–62

75. Buttice AL (2012) Aggregation of sediment and bacteria with mucilage from the *Opuntia ficus-indica* cactus. Dissertation, University of South Florida, Tampa

76. Renault F, Sancey B, Badot P, Crini G (2009) Chitosan for coagulation/flocculation processes-an ecofriendly approach. Eur Polym J 45(5):1337–1348. doi:10.1016/j.eurpolymj. 2008.12.027

77. De Souza MTF, Ambrosio E, De Almeida CA, De Souza Freitas TKF, Santos LB, de Cinque Almeida V, Garcia JC (2014) The use of a natural coagulants (*Opuntia ficus-indica*) in the removal for organic materials of textile effluents. Environ Moint Asses 186(8):5261–5271. doi:10.1007/s10661-014-3775-9

78. Bustillos LGT, Carpinteyro-urban S, Orozco C (2013) Production and characterization of *Opuntia ficus-indica* mucilage and its use as coagulant-flocculant aid for industrial wastewaters. Int J Biotechnol Res 1:38–45

79. Cardenas A, Goycoolea FM, Rinaudo M (2008) On the gelling behaviour of "nopal" (*Opuntia ficus-indica*) low methoxyl pectin. Carbohydr Polym 73(2):212–222. doi:10.1016/j.carbpol. 2007.11.017

80. Medina-Torres L, Vernon-Carter EJ, Gallegos-Infante JA, Rocha-Guzman NE, Herrera-Valencia EE, Calderas F, Jiménez-Alvarado R (2011) Study of the antioxidant properties of extracts obtained from nopal Cactus (*Opuntia ficus indica*) cladodes after convective drying. J Sci Food Agric 91(6):1001–1005. doi:10.1002/jsfa.4271

81. Torres LG, Carpinteyro-Urban SL, Vaca M (2012) Use of *Prosopis laevigata* seed gum and *Opuntia ficus-indica* mucilage for the treatment of municipal wastewaters by coagulation-flocculation. Nat Resour 3(2):35–41. doi:10.4236/nr.2012.32006

82. Miller SM, Fugate EJ, Craver VO, Smith JA, Zimmerman JB (2008) Toward understanding the efficacy and mechanism of *Opuntia* spp. as a natural coagulant for potential application in water treatment. Environ Sci Technol 42(12):4274–4279. doi:10.1021/es7025054

83. Pichler T, Young K, Alcantar N (2012) Eliminating turbidity in drinking water using the mucilage of a common Cactus. Water Sci Technol Water Supply 12(2):179–186. doi:10.2166/ ws.2012.126